# INTERNATIONAL CENTRE FOR MECHANICAL SCIENCES

COURSES AND LECTURES No. 93

J. LITWINISZYN

MINING COLLEGE, CRACOW

# STOCHASTIC METHODS IN MECHANICS OF GRANULAR BODIES

COURSE HELD AT THE DEPARTMENT
OF GENERAL MECHANICS
OCTOBER 1972

UDINE 1974

SPRINGER-VERLAG WIEN GMBH

ISBN 978-3-211-81310-2          ISBN 978-3-7091-2836-7 (eBook)

DOI 10.1007/978-3-7091-2836-7

# P R E F A C E

When discussing the mechanics of soil, rocks and loose media the models of the so called mechanics of continuous media are in general use. This model assumes the invariant of the contact relations between the elements of the media. In case of the above media being in motion the invariant relation of the contacts is not maintained. Contacts between these elements change, the ordered relation is not maintained, and the elements intermingle. The motion of the medium is characterized by the mass character of random changes in contact relations and consequently by random displacement of the medium elements.

The movement of such a collection of elements depends on their mechanical properties only in a small degree, being mainly dependent on their spatial structure. Since the interaction of the elements has a mass and random character, the summary effect of displacements of elements is defined by random laws in agreement with the central limiting theorems.

These heuristic considerations suggest the idea of describing the displacements of a loose medium on the basis of a model different from the model of a model different from the model of a continuous medium.

*That model may be regarded as a system of integral e-
quations which are generalizations of the Smoluchowski
equation describing the stochastic processes of the
Markov type. In particular, from this system a parabolic system of differential equations, defining the
mean values of displacement components of a loose medium, can be obtained.*

*Solutions for a number of cases of boundary
conditions of this system have been given. The results
have been compared with the displacement measurements
obtained in experiments carried out in a loose medium
in which the corresponding boundary conditions have
been realized.*

*J. Litwiniszyn*

*Udine, October 1971*

# STOCHASTIC METHODS IN MECHANICS OF GRANULAR BODIES

The mechanical phenomena in so called continuous media explained by means of a model based on the concept of continuous include phenomena for the explanation of which a continuous model is inadequate. In some cases we may feel that the mathematical model by means of which we describe the phenomenon is continuous, whereas the actual phenomenon described by the model is not continuous. The concept of noncontinuity seems to be inherent in the world of events and unavoidable.

The opposition of these two types of models of media, based on the model of continuum and model of a discrete medium, is known from the beginning of the history of mechanics.

Trials of reconciling these two opposed points of view involve basic considerations on the set theory and evolution of the concept of continuum. However, the unification of these two points of view continues to be an open problem.

The procedure used by Lagrange to derive the equation of vibrating strings is a good example of the trials of unifying these two points of view. Lagrange considered an arranged collection of $N$ points, the motion of each of which is described by a function differentiable according to time. This leads to a system of difference – differential equations, to which the limiting transition for $N \rightarrow \infty$ applies. Such a function requires ap-

propriate regularity of the findings describing the material coordinates of the points of the medium.

The assumption of regularity of these functions
imposes limits to the possibility of motion of the system. These
limits depend on the assumption of a contact relation between the
elements of the system, and consequently arrangement of the elements, the relation being an invariant during motion of the system.

The limitation following from such an assumption
in many cases leads to qualitative discrepancies between the rep-
resentation by the mathematical model and reality.

The limited class of admissible motions of the
medium obviously does not apply to the phenomena of motion of rar
efied gases. The phenomena of turbulent flow is  another example.

As is known, L.F. Richardson expressed doubts con
cerning the term wind velocity, i.e. whether the function describ
ing the coordinates of flowing elements of the medium are differ
entiable according to time. In the case of Brownian movements,
the measure, in the sense used by Wiener, of the set of differentiable functions describing the movements of the diffusing par
ticles is equal to zero. This means that nearly every trajectory
of particles exhibiting Brownian movements is undifferentiable.

Invariance of contact relations is not maintained
in the flow of fluids in porous media. On the whole, granular
bodies in motion do not fulfil this relation. Two grains of gran

ular medium lying in contact may separate after a brief period.
In that case, the condition of topologic transformation is not
fulfilled.

The phenomena described above characterize a geo̱
metric property of the collection of elements forming the medium,
namely contact relation of the elements. The continuity of the
medium is characterized by this relation. The medium is continuous if it cannot be divided into noncontacting parts.

Movement of a continuous medium is described by
a group of topologic trasformations with unchanging contact relations. In other words, during movement no new contacts are
formed, and existing contacts are not destroyed. This phenomenon
may be described as follows: Let a    be the Lagrangian coordinate of the medium. Movement of the medium with reference to an
immobile system of coordinates $\{x\}$ is described by the relation
$x = f(a,t)$   where    is time, and $a = f(a,0)$ is fulfilled.

The medium fulfils the condition of continuity
during motion if for each value of the continuous function $\epsilon =$
$= \epsilon(t) \geqslant 0$   there exists a number $\delta \geqslant 0$ so that the condition
$\varrho(a,b) < \delta$   is fulfilled only for each point b: then $\varrho[f(a,t),$
$f(b,t)] < \epsilon(t)$, where $\varrho = \varrho(\alpha,\beta)$ is the distance between points $\alpha$
and $\beta$ . In other words, if the distance $\varrho[(a,0),f(b,0)] = \varrho(a,b)$ is
not "very large", then    $\varrho[f(a,t),f(b,t)]$ also is "not very large".

Is this condition fulfilled by all media in mo-
tion? It can easily be demonstrated that it is not. The condition

is not fulfilled by granular media, in which the contact relation of the grains is changeable. As an example let us take a
granular medium consisting of a collection of sand grains. During
motion of the medium, the grains do not maintain this relation
invariably. Two grains lying in contact at a given moment, may
lose that contact.

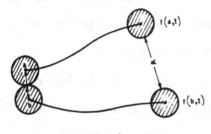

Fig. 1

This situation is illustrated in
Fig. 1. However small $\delta$ is, for
$\varrho(a,b) < \delta$ there is $\varrho[f(a,t),$
$f(b,t)] \geqslant \alpha$ hence not "every" val_
ue of $\varepsilon(t)$ can be taken, because
$\varepsilon(t) \geqslant \alpha$ must be fulfilled.

      The last examples indicate that the condition of
continuity is not maintained. In this case, application of the
methods of mechanics of continuous media is inadequate. Never-
the less, these methods are widely used, e.g. in soil mechanics,
or in classic mechanics of granular media. For the interpretation
of this category of phenomena it is reasonable to use a different
model, namely one in which the medium is regarded as a collection
of discrete elements, the contact relations of which are not main_
tained during motion. Contacts between these elements change, the
ordered relation is not maintained, and the elements intermingle.
We have here a phenomenon similar to that which occurs in fluid
or gas particles flowing turbulently, or performing Brownian mo_
tion.

However, the analogy between the motion of the elements of a granular medium with these phenomena is not complete. During the motion of granular media the freedom of movement of the elements is limited compared, for instance, with molecules of a rarefied gas, which has greater freedom.

2. Heuristic models based on the concept of random walk.

Motion of a granular medium is characterized by the mass character of the random changes in contact relations, and consequently random displacement of grains.

Hence, it is reasonable to regard the motion of a mass of a granular medium as random process. [1]

## MODEL I

For preliminary heuristic considerations on the motion of a granular medium as a random process, let us imagine a system of cages illustrated in Fig. 2.     Each cage contains a ball subjected to the force of gravity. Let the system of cages fulfil the condition that removal of a ball from the horizontal in the second layer stratum to take its place, assuming equal probability of both events, i. e. 1/2.

As a result of the displacement of the ball from cage $a_2$ to $a_1$ , cage $a_2$ will be occupied by a ball from cage $a_3$

or $b_3$ from layer III. Similarly, removal of the ball from cage $b_2$ will cause its place to be taken by a ball from cage $b_3$ or $c_3$ .

Removal of a ball from cage $a_1$ empties one of the cages $a_3$ , $b_3$ or $c_3$ , the probability of these events being 1/4, 2/4 and 1/4 respectively. The distribution of the probabilities of these events in cages $a_4$ , $b_4$ , $c_4$ and $d_4$ in layer IV will be 1/8, 3/8 and 1/8.

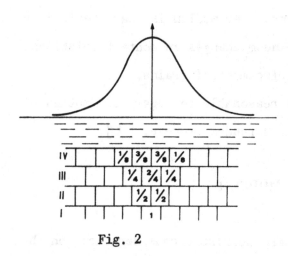

Fig. 2

The distribution of probabilities is illustrated in Fig. 2. Instead of one, a larger number of balls is removed from cage $a_1$ , then the cages in the highest layers which formerly contained balls will be emptied.

The boundary of the emptied cages forms a step-wise line. Given a sufficiently dense network of cages, removal of a sufficient number of balls from cage $a_1$ will give a step-wise line approaching Gauss's K curve, symmetrical with respect to a perpendicular straight line passing through the centre of cage $a_1$ .

The result described above, consisting in the

formation of a contour of emptied cages in the upper layers of
the system was predicted exclusively on the basis of elementary
probability calculus, similarly to the prediction of the frequen
cy of a coin cast into the air falling heads or tails up, or of
drawing a given playing card from a deck.

In these predictions it is not the individual
properties of the balls that count, but the structure of their
collection.

Although the procedure outlined above does not
permit description of the fate of individual balls, (given that
a sufficiently large number of balls has been removed from the
cage $a_i$ ) it does make it possible to predict some of the properties of the collection of balls (e.g. the law of distribution
of the emptied cages).

The so called "limit theorems" of the theory of
probability allow us to prognosticate. The practical importance
of these theorems is that they show that mass random phenomena
are governed by strict, not random, regularities. In other words,
mass random events lead to a certain degree of regularity.

The choice of statistical methods as opposed to
deterministic methods may be regarded either as an attempt to
avoid conceptual and analytical difficulties, or as a desire to
describe reality more accurately.

Regardless of the approach, however, the mathematical implications are the same.

For a more precise description of the above model,
let us imagine a plane system of cages in the Cartesian system
of coordinates $\{x,z\}$ with the $z$ axis directed perpendicularly
upward, as shown in Fig.3.
Each cage contains material particles subject to the force of
gravity acting in parallel
to the $z$ axis but in the
opposite direction. Under
the influence of gravity,
the particles can only
drop downward.

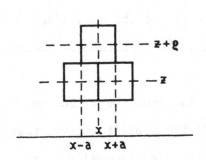

Fig. 3

Let us consider the system
of cages illustrated in the figure, assuming that material particles can be displaced only toward lower laying cages (e.g.
from $C$ to $A$ or $B$ ).

The downward migration of particles corresponds
with an oppositely directed migration of empty spaces. Downward
migration of material particles and corresponding upward migration of voids are two attributes of motion of a granular medium.

Let us assume that the random migration of voids
in the system of cages illustrated in Fig. 3 takes place as fol
lows.

A void migrates from cage $A$ to $C$ , or from $B$
to $C$ . Let the probability of these two events be $p$ and $q$
respectively, assuming that $p+q=1$. To cages $A$ , $B$ and $C$

may be assigned the corresponding coordinates (Fig.3).

$$(x - a, z), (x + a, z), (x, z + \varrho) .$$

Let $P = P(x,z)$ denote the probability of the occurrence of an empty cage with the coordinate $(x,z)$ .

In accordance with the mechanism of random migra
tion of voids we may then write

$$P(x, z + \varrho) \;=\; p P(x - a, z) + q P(x + a, z) .$$

Applying Taylor's development to the terms in the
last equation, and considering that $p + q = 1$ , we get:

$$\varrho \frac{\partial P(x,z)}{\partial z} \;=\; a(q - p) \frac{\partial P(x,z)}{\partial x} + \frac{a^2}{2} \frac{\partial P(x,z)}{\partial x^2} + \dots . \qquad (2.1)$$

Proceeding in the last equation to the limit for
$\varrho \longrightarrow 0$ and $a \longrightarrow 0$ we must do so in such a manner that the expect
ed value of the displacement of the void in the direction of the
$x$ axis and variance of the displacement will be finite values
for all $z > 0$. The expected value of displacement of the void for
the given coordinate $z$ is $(p - q) a \frac{z}{\varrho}$ , and the respective vari
ance is

$$\left[ 1 - (p - q)^2 \right] a^2 \frac{z}{\varrho} \;=\; 4 p q z \frac{a^2}{\varrho} .$$

Finiteness of the variance requires $a^2/\varrho$  to be limited. Finiteness of the expected value requires $(p-q)$ to be of the same

order as **a** .

Assuming such an order of magnitude, we may deter-
mine the limits:

$$(2.2) \qquad A = \lim_{\substack{a \to 0 \\ \varrho \to 0}} \frac{1}{2}\frac{a^2}{\varrho} , \quad B = -\lim_{\substack{a \to 0 \\ \varrho \to 0 \\ p \to q}} \frac{a(q-p)}{\varrho} .$$

After the limit process in equation 2.1 for
$a \to 0, \varrho \to 0$ we get

$$(2.3) \qquad \frac{\partial P}{\partial z} = A \frac{\partial^2 P}{\partial x^2} - B \frac{\partial P}{\partial x} .$$

This equation may be generalized by assuming that
A    and  B  are functions of the coordinate  z     .

Equation 2.3 then takes the form

$$(2.4) \qquad \frac{\partial P(x,z)}{\partial z} = A(z) \frac{\partial^2 P(x,z)}{\partial x^2} - B(z) \frac{\partial P(x,z)}{\partial x} .$$

The parabolic equation obtained is of the same
type as the equation describing thermal conduction or diffusion
with convection. This equation will be derived from postulates
constituting the basis for conduction or diffusion with convection. This equation will be derived from postulates constituting
the basis for construction of a more general model describing
motion of a granular medium.

The function $P = P(x,z)$ being a solution of the equation 2.4, describes the probability of the appearance of an empty

cage assigned to the coordinate $(x,z)$.

If a particle is removed from a cage with the co
ordinate $(x_0, z_0)$ the void will be filled by another particle which
under the influence of the force of gravity falls into they empty
space from the overlying area. The phenomenon of random walk of
the particle and random walk of the void correspond.

If not merely one, but a larger number of materi
al particles are removed from the cage $(x_0, z_0)$ , the voids will
be distributed at random in the area $z > z_0$ .
If the number of particles removed from the cage $(x_0 z_0)$ is sufficiently large, then $P = P(x,z)$ may be regarded as the approximate
value of the ratio of the volumes of the voids which have migrat
ed through the cage with the coordinate $(x,z)$ to the volumes of
the voids in the cage with the coordinate $(x,z)$ .

In other words, $P = P(x,z)$ is approximately propor
tional to the volume of the voids which migrated through cage
$(x,z)$        . In accordance with the law of great numbers, the
approximation is better if the number of voids is larger.

If the number of voids migrating through cage
$(x, z)$ is sufficiently large, the volume of the voids can be
measured.

Let us imagine horizontal coloured layers in the
granular medium formed by grains of the medium. As a result of
the displacements, the original coloured layers will form troughs.
The depth of these troughs was measured, and the results compared

with the solutions of equation 2.3. The experiment in which this
measurement was made was carried out under the following conditions. A vertical box in the shape of a narrow rectangular par-
allelepiped was filled with sand. The sand was observed through
the glass side walls of the box. These aluminium wires were placed
in the sand perpendicularly to these walls and the coordinates
of the axis of the wires were measured.

Sand emerged from the bottom of the box through
a narrow orifice S (Fig. 4). After a certain volume of sand had
flowed out of the box, a field of displacements of the sand was
formed, which measured on the basis of the displacement of the
wires in the sand.

Let us assume a system of $(x, z)$ coordinates as in
Fig. 4 the axis being directed vertically upward. This direction
of the axis is chosen on the assumption that the displacements
take place under the influence of gravity.

The magnitude of the displacements, as described
above, is proportional to $P = P(x, z)$. The state of the displacements is caused by the outflow of sand through the orifice S .
Let us assume that the length $AC$ is sufficiently large and the
influence of the walls $AB$ and $CD$ on the displacement within
the box is imperceptible.

The condition given by the outflow of a certain
volume of sand through the orifice S in the bottom of the box
for $z = 0$, $x = 0$, can be described by the relation:

$$P(x,0) = \gamma \delta(x) \qquad (2.5)$$

where $\delta = \delta(x)$ is the so-called "function" of Dirac, and $\gamma$ is the volume of grains of the granular medium flowing from the narrow aperture S .

The solution of equation 2.3 under the initial condition 2.5 has the form

$$P(x,z) = \gamma(4\pi Az)^{-\frac{1}{2}} \exp\left[-\frac{(x-Bz)^2}{4Az}\right]. \qquad (2.6)$$

If the number of grains removed at point $x = z = 0$ is sufficiently large, then for the constant value $z$ (2.6) describes the shape of the trough formed under the condition 2.5.

As may easily be shown

$$\int_{-\infty}^{+\infty} P(x,z)dx = \gamma .$$

Hence, the volume of the trough is independent of $z$ and equal to the volume $\gamma$ of granular medium thrown out through the orifice S . The solution 2.6 was compared with the results of the measurements of displacements under the initial conditions in the form of 2.5.

Experiments were carried out with two types of media. In the first experiment, an isotropic medium was used in the form of sand with narrow fraction of grains. In the second experiment an anisotropic medium consisting of sand with artifi cially produced anisotropy. Anisotropy was produced by placing

mica in uniform parallel planes in the granular medium. Such a
medium is anisotropic because it has different properties in va
rious directions.

In the isotropic medium, the probability of ran-
dom walk of a particle or void in the direction of the $x$ axis
is identical. Hence $p=q=\frac{1}{2}$ . In this case, according to equation
2.2, $B=0$ . The solution 2.6 then takes the form

$$(2.7) \qquad P(x,z) \;=\; \gamma(4\pi A z)^{-\frac{1}{2}} \exp\left[-\frac{x^2}{4Az}\right].$$

If the number of grains removed through the orifice
S  is sufficiently large, then $P = P(x,z)$ describes the shape of
the trough in the granular medium. A comparison of the measurements of the troughs with solution 2.7 for $z=\text{const.}$ is shown in
the diagram Fig. 4.

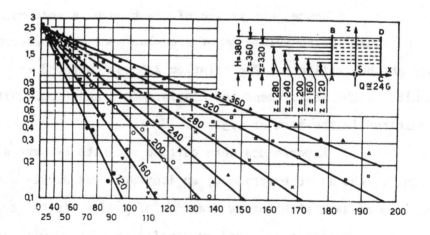

Fig. 4

The function scale in the diagram is such that the symmetric
halves of Gaussian curves described by equation 2.7, are transformed into halves-straight. The halves-straight in diagram
are straightened out symmetric half-curves, and the points plot
ted on them are the results of measurements.

In the second experiment, conditions were realiz
ed under which $p \neq q$ , i.e. the probability of random walk of the
void in a direction parallel to the $x$ axis was not symmetric.
This effect was obtained when the displacements were accomplish
ed in an anisotropic granular medium as described above. The
troughs formed in this medium are described by the equation 2.6.

According to this solution, troughs are formed in
the mass of the granular medium which, similarly to the case
described by equation 2.7, have the bell-shape of Gaussian curves.
However, with increasing height of $z$ , they deviate in the direc
tion of the $x$ axis. The magnitude of the deviation in relation
to the vertical axis passing through point $x = 0$ , $z = 0$ (the point
at which the initial conditions of Dirac were given) 2.5 is defined by $Bz$ from equation 2.6.

Hence, the phenomenon of deviation is dependent
on coefficient $B$ in equation 2.3.

Coefficients $A$ and $B$ in equations 2.3, 2.4 can be
given physical intrepretations. Coefficients $A$ defines dispersion
characterizing the Gaussian subsidence trough, and coefficient $B$
characterizes ordered movement of the particles connected with

anisotropy of the medium.

The results of experiments are presented in Fig. 5, in which the inclined layers of the trough, by means of which

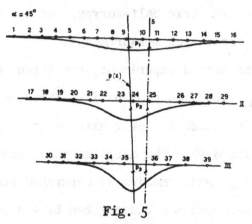

Fig. 5

anisotropy of the medium was accomplished, are shown. The troughs formed as a result of the evacuation of the sand were measured at fixed levels $z$ .

Fig. 5 shows these troughs at different levels I, II and III which in accordance with the theoretical results are subject to deviation $\varrho = \beta z$ . The figure pertains to the experiment in which the angle of inclination of the layers of the trough to the horizontal plane was $45^{\circ}$.

The profiles of the troughs obtained in the experiment at levels I, II and III were plotted on functional diagrams, in which symmetrical halves of Gaussian curves are transformed into halves-straight.

Fig. 5 represents the points of measured coordinates of depth of the measured troughs at levels I, II, III plot

ted on the functional diagram, and straight lines representing
transformed Gaussian curves depicting the theoretical shape of
the troughs.

Generalization of equation 2.3 to the form 2.4
takes into account the lack of homogeneity of the medium. The
dependence of coefficients $A$ and $B$ takes into account changes
in the properties of the medium from this variable. Solution of
equation 2.4 for the initial condition 2.5 has the form:

$$P(x,z) = \gamma[4\pi\zeta(z)]^{-\frac{1}{2}} \exp\left\{-\frac{[x - \varrho(z)]^2}{4\zeta(z)}\right\}$$

where

$$\varrho(z) = \int_0^z B(s)ds , \quad \zeta(z) = \int_0^z A(s)ds .$$

For the established magnitude of coordinate $z$ we
again get the function $P = P(x,z)$ in the form of a Gaussian curve,
subject to deviation in the direction of the $x$ axis, with the
magnitude $\varrho = \varrho(z)$ in relation to the vertical axis passing
through point $x = 0, z = 0$.

Besides experiments in which the initial condi-
tions to equation 2.5, an experiment was performed in which this
condition was realized in the form:

$$P(x,0) = \begin{cases} 0 & \text{for} \quad x < 0 \\ \omega_0 = \text{const.} & \text{for} \quad x \geq 0 . \end{cases} \tag{2.8}$$

This experiment was carried out in a horizontally homogeneous medium in which the process of the formation of the subsidence trough is described by equation 2.4.

The solution of this equation for the initial condition 2.8 takes the form:

$$(2.9) \qquad P(x,0) = \frac{\omega_0}{\sqrt{\pi}} \int\limits_{-\frac{1}{2}x\left[\int\limits_0^z A(s)ds\right]}^{\infty} e^{-\lambda^2} d\lambda \ .$$

In agreement with equation 2.9, for established values of the troughs produced by initial conditions 2.8 take the form erf.

The results of measurements in the experiment on the shaping of the trough in granular medium produced by conditions 2.8 are represented by Fig. 6, using a functional scale in

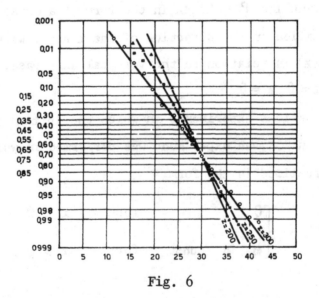

Fig. 6

which the pattern of erf is transformed into a straight line.
The plotted points are results of measurement, and the straight
lines represent erf, which is a theoretical solution.

Another test of solution 2.9 consists in the com
parison of this solution with the curves representing troughs
produced on the surface of the earth's crust as a result of sub
terranean exploitation of useful minerals.

For the purpose of this confrontation, cases were
selected in which the displacements could be assumed to be approximately flat, and conditions of exploitation could be described by equation 2.8.

Eight such troughs which approximately fulfilled
the above conditions were selected from nature.

The result of measurements, converted into nondimensional magnitudes, were represented by points plotted on
Fig. 7. On the vertical axis, the nondimensional magnitude,
$\dfrac{P}{\omega_{max}}$ was plotted, in which $\omega_{max}$ is equal to the magnitude

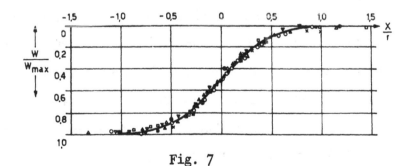

Fig. 7

$\omega_0$  in the initial condition 2.8 (these values varied in the different troughs under consideration). On the horizontal axis, the nondimensional magnitude $\frac{x}{r}$ was plotted, where $r$ is a parameter with length dimension. The continuous curve in the figure represents the erf. The collection of points representing the results of measurements grouped near the curve erf is interest→ ing, indicating that the course of the phenomenon very nearly agrees with equation 2.9.

However, this reaction, being an implication of the differential equation 2.4 and initial conditions 2.8, does not contain any material constants such as those in equations describing notion of continuous media. The equation giving the relation between the state of deformation and stress has not been used, as this concept was not introduced in our considerations. How, then, can this interesting agreement between the theoretical and observed results be explained?

The movement of the earth's crust is caused by the removal of rock masses. As a result of this movement, an area of rock mass lying above the removed volume is crushed, giving rise to macrorubble, which is not a continuous medium. This macrorub ble may be regarded as a discrete medium, consisting of a collec tion of many elements, which during their random walk act one upon another, as, for example happens during movement of sand.

The movement of such a collection of elements de pends only in a small degree on their mechanical properties, be

ing mainly dependent on their spatial structure, e.g. inhomoge-
neity or anisotropy of the position of the elements of the collection. Since the interaction of the elements has a mass and
random character, the summary effect, consisting in formation
of the trough as a result of slight displacements of individual
elements, in agreement with the central limiting theorem, is
determined by normal distribution. In the case of boundary conditions in the form of 2.8, the effect is described by the for-
mula which is a result of the superposition of normal distribution.

These facts can explain the phenomenon, found
experimentally in the laboratory or observed in nature, of the
shaping of the trough in agreement with the law of normal distri
bution, regardless of the mechanical properties of the rocks
forming the macrorubble in which the trough is shaped. In the
case under consideration, the statistical method provides infor-
mation about the movement of the collection of elements without
a knowledge of their individual fates.

The above heuristic considerations of a probabilistic character lead to the linear equation of parabolic type,
or so-called Fekker-Planck equation.

Among others, the phenomenon of termal conduction
and diffusion are governed by equations of this type. Mathematical analogy exists between these phenomena and the phenomena of
displacement in a granular medium governed by the equation 2.3,

**2.4.**

In the parabolic equation governing thermal conduction or diffusion, time coordinate corresponds to the $z$ coordinate in equations 2.3 or 2.4.

The solution of these equations is characterized
by the fact that a state of temperature or concentration induced
in any point of the medium in which the phenomenon takes place
is manifested at once in every other point of the medium. This
is a fundamental difference between these phenomena described
by hyperbolic equations. Phenomena described by the latter equa
tions are characterized by finite velocity of propagation of the
wave.

In the case of displacement of granular media,
the afore-mentioned property of parabolic equations describing
probability of occurrence of displacements is manifested as fol
lows. Displacements given at point $(x_0 z_0)$ appear for $z > z_0$ through
out the whole interval $x$ , i.e. for $-\infty > x < +\infty$ .

Another qualitative property of the solution of
parabolic equations is that discontinuities given under initial
conditions at time $t = t_0$ , or in our case at the level $z = z_0$ ,
in the area $z > z_0$ are "washed out".

In contrast to parabolic equations, the solutions
of hyperbolic equations are characterized by the fact that the
discontinuities given in the initial conditions are not "washed
out", but are propagated along the so-called characteristics of

these equations.

The question arises as to which of the afore-
mentioned properties of the equations corresponds to the actual
phenomenon of displacement in a granular medium. Depending on
the actual experimental conditions, phenomena corresponding
either to parabolic or to hyperbolic equations may be observed.

This, in turn, raises the question as to whether
probabilistic considerations leading to parabolic equations can
also give a mathematical model described by a hyperbolic equation.

As a matter of fact, this is possible, and Goldstein has described such a model. [ 2 ]

## MODEL II [ 3 ]

In order to sketch this model, let us consider
a plane system of cages in the Cartesian system of coordinates
$\{x,z\}$ with the $z$ axis directed perpendicularly upward. As be-
fore, we assume that the material particles in the cages can
drop into one of two lower-lying cages. Downward migration of
particles corresponds to migration of voids in the opposite direction, i.e. upwards.

The previously considered model depicted the
state of displacement in a granular medium independent of time,
i.e. an asymptotic state established after a sufficiently long

period of time after the initial conditions were given.

   We may now generalize our considerations of the
phenomenon of displacement in time and space.

The system of cages illustrat‐
ed in 'Fig. 8   may be used
as a case in point. Migration
of voids in this system of
cages takes place as follows.
A void created in cage A
can migrate through cages A
C , D or through A , C ,
E    . In the first case,
the direction of the second
step of the migrating void

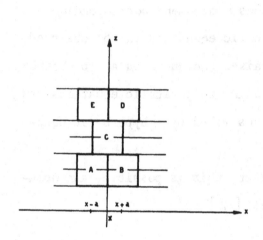

Fig. 8                     will be same as the direction
of the second step of the migrating void will be the same as the
direction of the first step. In the second case, the direction
will change.

   Respectively, a void in cage B can migrate
through cages B , C , E or through B , C , D .

   Let p be the probability of a void migrating
in the same direction during two consecutive steps.

   The magnitude $q = 1 - p$ is the probability of
the void changing its direction.

   We introduce the magnitude $C = p - q$ .

Let $P = P(x,z,t)$ be
the probability of a void appearing
in a cage with the coordinates $(x,z)$
at time $t$ .

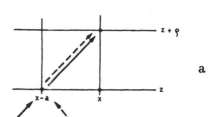

$\alpha(x,z,t)$     — probability of a
void in cage $(x,z)$
migrating down-
ward in the di-
rection from left
to right at time $t$,

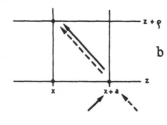

Fig. 9

$\beta(x,z,t)$     — probability of a void in cage $(x,y)$   migrating
downward in the direction from right to left at
time $t$ .

The relation exists

$$P(x,z,t) \;=\; \alpha(x,z,t) + \beta(x,z,t) \;.$$

The relation may be written (Fig. 9a)

$$\alpha(x,z + \varrho, t + \tau) \;=\; p\alpha(x - a, z, t) + q\beta(x - a, z, t) \qquad (2.10)$$

where $\tau$ is the period of time during which the void passes
from level $z$ to level $z + \varrho$ .

From the last two relations it follows that:

$$\alpha(x, z + \varrho, t + \tau) \;=\; pP(x - a, z, t) - c\beta(x - a, z, t) \qquad (2.11)$$

where $c = p - q$

and analogously  Fig. 9b

$$(2.12) \qquad \beta(x,z + \varrho, t + \tau) \;=\; p\beta(x + a,z,t) + q\alpha(x + a,z,t)$$

$$(2.13) \qquad \beta(x,z + \varrho, t + \tau) \;=\; pP(x + a,z,t) - c\alpha(x + a,z,t).$$

Equations 2.10 and 2.12 can be written in the form

$$\alpha(x + a,z,t) \;=\; p\alpha(x, z - \varrho, t - \tau) + q\beta(x,z - \varrho, t - \tau)$$

$$\beta(x - a,z,t) \;=\; p\beta(x, z - \varrho, t - \tau) + q\alpha(x,z - \varrho, t - \tau).$$

adding the last equations on each side, we get

$$\alpha(x + a,z,t) + \beta(x - a,z,t) \;=\; p\alpha(x,z - \varrho, t - \tau) + q\beta(x,z - \varrho, t - \tau) +$$
$$+ p\beta(x,z - \varrho, t - \tau) + q\alpha(x,z - \varrho, t - \tau) \;=\;$$
$$=\; P(x,z - \varrho, t - \tau).$$

Adding the sides of equations 2.11, 2.13 gives

$$\alpha(x,z + \varrho, t + \tau) + \beta(x,z + \varrho, t + \tau) \;=\;$$
$$=\; p\left[P(x - a,z,t) + P(x + a,z,t)\right] - c\left[\alpha(x + a,z,t) + \beta(x - a,z,t)\right]$$

and finally

$$(2.14)\; P(x,z + \varrho, t + \tau) \;=\; p\left[P(x - a,z,t) + P(x + a,z,t)\right] - cP(x,z - \varrho, t - \tau).$$

This difference equation describes the probability of propagation of voids in a model representing a granular medium regarded as a collection of discrete elements. At the same time, equation 2.14 is the starting point for obtaining the differential

equations, assuming existence of appropriate limits.

By developing function $P$ around $(x,z,t)$ into a series, we get

$$P(x,z+\varrho,t+\tau) = P(x,z,t) + \frac{\partial P}{\partial z}\varrho + \frac{\partial^2 P}{\partial z^2}\frac{\varrho^2}{2} + \frac{\partial P}{\partial t}\tau + \frac{\partial^2 P}{\partial t^2}\frac{\tau^2}{2} + \frac{\partial^2 P}{\partial z \partial t}\varrho\tau + \ldots$$

$$P(x-a,z,t) = P(x,z,t) - \frac{\partial P}{\partial x}a + \frac{\partial^2 P}{\partial x^2}\frac{a^2}{2} + \ldots$$

$$P(x+a,z,t) = P(x,z,t) + \frac{\partial P}{\partial x}e + \frac{\partial^2 P}{\partial x^2}\frac{a^2}{2} + \ldots$$

$$P(x,z-\varrho,t-\tau) = P(x,z,t) - \frac{\partial P}{\partial z}\varrho + \frac{\partial^2 P}{\partial z^2}\frac{\varrho^2}{2} - \frac{\partial P}{\partial t}\tau + \frac{\partial^2 P}{\partial t^2}\frac{\tau^2}{2} + \frac{\partial^2 P}{\partial z \partial t}\varrho\tau + \ldots$$

By introducing these magnitudes into equation 2.14, we get:

$$P + \frac{\partial P}{\partial z}\varrho + \frac{\partial^2 P}{\partial z^2}\frac{\varrho^2}{2} + \frac{\partial P}{\partial t}\tau + \frac{\partial^2 P}{\partial t^2}\frac{\tau^2}{2} + \frac{\partial^2 P}{\partial z \partial t}\varrho\tau + \ldots \; =$$

$$= \; p\left[ P - \frac{\partial P}{\partial x}a + \frac{\partial^2 P}{\partial x^2}\frac{a^2}{2} + \ldots + P + \frac{\partial P}{\partial x}a + \frac{\partial^2 P}{\partial x^2}\frac{a^2}{2} + \ldots \right] +$$

$$- c\left[ P - \frac{\partial P}{\partial x}\varrho + \frac{\partial^2 P}{\partial z^2}\frac{\varrho^2}{2} - \frac{\partial P}{\partial t}\tau + \frac{\partial^2 P}{\partial^2 t}\frac{\tau^2}{2} + \frac{\partial^2 P}{\partial z \partial t}\varrho\tau + \ldots \right]$$

and by ordering the expressions in the last equation, thus the result

$$P(1 - 2p + c) + \frac{\partial P}{\partial z}\varrho(1 - c) + \frac{\partial^2 P}{\partial z^2}\frac{\varrho^2}{2}(1 + c) + \frac{\partial P}{\partial t}\tau(1 - c) + \qquad a$$

$$+ \frac{\partial^2 P}{\partial t^2} \frac{\tau^2}{2}(1+c) + \frac{\partial^2 P}{\partial z \partial t} \varrho\tau(1+c) - \frac{\partial^2 P}{\partial x^2} p \frac{a^2}{2} + \dots = 0$$

is obtained, which in turn, because:

$$1 - 2p + c = 1 - 2p + p - q = 1 - (p+q) = 0.$$

Hence the last equation may be written:

(2.15)
$$\frac{\partial^2 P}{\partial z^2} + \frac{\partial^2 P}{\partial z \partial t} 2 \frac{\tau}{\varrho} + \frac{\partial^2 P}{\partial t^2} \frac{\tau^2}{\varrho^2} - \frac{\partial^2 P}{\partial x^2} \frac{a^2}{\varrho^2} \frac{p}{1+c} +$$

$$+ 2 \frac{\partial P}{\partial z} \frac{1-c}{\varrho} \frac{1}{1+c} + 2 \frac{\partial P}{\partial t} \frac{\tau(1+c)}{\varrho^2} \frac{1}{1+c} + \dots = 0.$$

In the last equation, we pass to the limit

for

$$\varrho \to 0, \quad a \to 0, \quad \tau \to 0, \quad c \to 1$$

in the case when $c \to 1$ , then $p \to 1$ , because $p + q = 1$ ,
$p - q = C$ and if $c = 1$ , then $p = 1$ .

Let us assume the following limits:

$$\lim_{\substack{\varrho \to 0 \\ \tau \to 0}} \frac{\varrho}{\tau} = A > 0; \quad \lim_{\substack{a \to 0 \\ \varrho \to 0 \\ \tau \to 0}} \frac{a^2}{\varrho\tau} = B > 0; \quad \lim_{\substack{c \to 1 \\ \tau \to 0}} \frac{1-c}{\tau} = D > 0.$$

From this assumption it follows that:

$$\frac{D}{A} = \frac{1-c}{\varrho}; \quad \frac{D}{A^2} = \frac{\tau(1-c)}{\varrho^2}; \quad \frac{B}{A} = \frac{a^2}{\varrho^2}.$$

Taking these relations into consideration, as
well as the fact that when passing to the limit, expressions of
higher order disappear, marked by dots in equation 2.15, we ob-

tain from this equation:

$$\frac{\partial^2 P}{\partial^{2}} + \frac{2}{A}\frac{\partial^2 P}{\partial z \partial t} + \frac{1}{A^2}\frac{\partial^2 P}{\partial t^2} - \frac{B}{2A}\frac{\partial^2 P}{\partial x^2} + \frac{D}{A}\frac{\partial P}{\partial z} + \frac{D}{A^2}\frac{\partial P}{\partial t} = 0 . \qquad (2.16)$$

Let us see what is the physical interpretation

of the function $P = P(x,z,t)$ .

If a time $t = t_0$ , a void appears at the point with the coordi-

nates $(x_0, z_0)$, which will be replaced by a material particle.

Since this phenomenon is reproducible, the void wanders random-

ly upward. The probability of a void appearing in a point with

the coordinates $(x,z)$ for $z > z_0$ at time $t > t_0$ is defined by the

function $P = P(x,z,t)$  . Let us imagine that not one, but a num

ber of material particles were removed from point $(x_0, z_0)$ in the

time interval $(t_0, t)$ . In that case, the resulting voids will

be scattered at random in the area $z > z_0$ in time $t > t_0$ .

If the number of particles removed is sufficient-

ly large, then $P = P(x,z,t)$ may be regarded as the ratio of the

volume of the voids which passed through point $(x,z)$ to the sum

of the volumes of particles removed from the point with coordi-

nates $(x_0, z_0)$ . In other words, the magnitude of $P = P(x,z,t)$ is

proportional to the sum of volumes of particles which migrated

through the horizontal surface element ascribed to point $(x,z)$

If a sufficiently long time has elapsed after the

establishment of the initial condition, we may consider that the

migration of voids in the granular medium has finished. In such

an asymptotic state, derivatives according to time disappear in equation 2.16, and the equation assumes the form:

$$(2.17) \qquad \frac{\partial P}{\partial z} + \frac{A}{D}\frac{\partial^2 P}{\partial z^2} = \frac{B}{2A}\frac{\partial^2 P}{\partial x^2} \ .$$

The last equation defines the probability $P = P(x,z)$ of a void appearing at point $(x,z)$ after an asymptotic state, sufficiently distant from the initial conditions, develops in the medium.

In order to solve equation 2.17 when the boundary conditions are given for $z = 0, -\infty < x < +\infty$, the values of $P = P(x,0)$ and $\left(\frac{\partial P}{\partial z}\right)_{z=0}$ must be known.

In the experiments that were carried out, the first condition was fulfilled by removing a given volume of the granular medium through the slit in the bottom of the box containing the medium. Physical realization of the second initial condition for the solution of the equation 2.17 is a less simple matter. We assume the second initial condition as an auxiliary hypothesis.

We therefore postulate conditions for $z = 0$ in the form:

$$(2.18) \qquad P(x,0) = \gamma \, \delta(x)$$

$$(2.19) \qquad \left(\frac{\partial P(x,z)}{\partial z}\right)_{z=0} = 0 \ .$$

It is evident that conditions 2.18 and 2.5 are identical, and that condition 2.19 is hypothetical.

Solution of equation 2.17 for condition 2.18 and 2.19 has the form:

$$P(x,z) = \delta \frac{D}{2}\left(\frac{A^2 B}{2D}\right)^{-\frac{1}{2}} \exp\left[-\frac{zD}{2A}\right]\left\{I_0(y) + \frac{zD}{2A}\frac{I_1(y)}{y}\right\} \quad \text{for} \quad |x| < \left(\frac{B}{2D}\right)^{\frac{1}{2}}z$$

and

$$P(x,z) \equiv 0 \quad \text{for} \quad |x| > \left(\frac{B}{2D}\right)^{\frac{1}{2}}z$$

(2.20)

where

$$y = \frac{1}{2}\left(\frac{A^2 B}{2D}\right)^{-\frac{1}{2}}\left(\frac{B}{2D}z^2 - x^2\right)^{\frac{1}{2}}$$

and $I_0$ and $I_1$ are Bessels functions of the second kind, determin ed from the relation:

$$I_0(y) = \sum_{m=0}^{\infty}\frac{\left(\frac{1}{2}y\right)^{2m}}{(m!)^2}; \quad I_1(y) = \sum_{m=0}^{\infty}\frac{\left(\frac{1}{2}y\right)^{2m+1}}{m!(m+1)!}$$

introducing the variables:

$$\xi = \frac{D}{2}\left(\frac{A^2 B}{2D}\right)^{-\frac{1}{2}}x; \quad \eta = \frac{1}{2}\frac{D}{A}z.$$

The solutions 2.20 of the equation 2.17 for boundary conditions 2.18, 2.19 are illustrated in Fig. 10 (see next page). This solu tion is a discontinuous one, and discontinuity occurs along the characterics of equation 2.17

$$x = \pm\left(\frac{B}{2D}\right)^{\frac{1}{2}}z.$$

The qualitative effect of solution 2.20 consisting in discontin uity of the characteristics was observed in experiments Fig. 11.

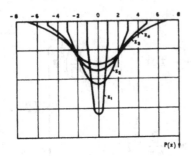

Fig. 10

The effect occurs in the lower layers of the granular medium near the slit in the bottom of the box. With the increasing of the distance from the points of the given boundary conditions, discontinuities disappear, and are no longer discernible in the upper troughs of the granular medium.

Troughs at higher levels, more distant from the level $z = 0$, have the shape of bell curves with a wide base.

The qualitative effects described above suggest change of the equation from hyperbolic to parabolic. However, the construction of such an equation is an open problem which still remains to be solved.

## 3. GENERAL METHOD OF DISPLACEMENTS OF A GRANULAR MEDIUM AND ITS POSTULATES [1,4]

The considerations presented above lead to linear differential equations of the parabolic (Fokker–Planck) or hyperbolic (telegraphic) type.

The equations describe the probability of voids appearing in the areas of granular medium under consideration. The solutions of these equations cannot be compared directly with observations.

However, if the number of voids in a given point

increases, then in agreement with the law of great numbers, the

ratio of the number of voids

to the number of voids given

in boundary conditions 2.5

approaches the calculated

probability.

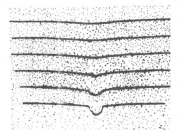

The volume of

Fig. 11

the voids at a given point can be measured by measuring the depth

of the trough ascribed to that point. Such measurements have

been made in laboratory experiments on displacements in granular

medium, and have also been observed in the earth's crust under-

going displacement during subterranean mining operations.

The results of these measurements and their con-

frontation with phenomena of moderately large displacements

prompted an attempt at a general conception and construction of

a model of the phenomenon based on a set of postulates *

---

* The need of such a generalization arises from the following
reasoning. Let equation 2.3 be given for $B = 0$, i.e. in the form:

$$\frac{\partial P(x, z)}{\partial z} = A \frac{\partial^2 P(x, z)}{\partial x^2}$$

As is known, if we assume in Cauchy's problem $P(x, 0) \equiv 0$ we have
$P(x, z) \equiv 0$ for $z > 0$.

On the other hand, the following experiment may be made. Let
us take a horizontal layer of a granular medium resting on two

The general idea is based upon the concept of displacement.

In order to arrive at this concept, let us consider surface area $F$ in Euclidean space with the measure $f$. The set of material elements of the medium displaced through $F$ will be designated by the symbol $W$. Let the set $W$ be measurable, its measure being $w$. Let us consider the function $w = w(P)$ for $P \in F$ and create limits in point $P_0 \in F$

$$\lim_{\substack{f \to P_0 \\ P \in F}} \frac{1}{F} \int_f w(P)df = w(P_0).$$

Then $w = w(P_0)$ is the limit of the ration of the volume of material displaced through $F$ to the measure $f$ tending towards $P_0$, $w = w(P)$ may be called the displacement averaged.

We now introduce the Cartesian coordinate system $\{x^1, x^2, x^3\}$. The components of the vector of displacement at point $P_0$ of this system will be designated.

$$w^i = w^i(P_0) \quad \text{for } i = 1,2,3.$$

---

plates. One of the plates is moved parallel to the other. This kind of movement only displaces the granules of the medium on the plates horizontally. Vertical displacement is equal to zero. In spite of the conditions on the surface of the plates, vertical displacements will occur in the upper layers of the granular medium. This is contrary to the Cauchy's solution for homogeneous initial conditions.

The above facts prompted attempt at generalization and construction of a modified model of the displacements in granular media.

The process of displacement of the granular me-
dium is defined in the magnitudes $w^i$ and all the points of the
space under consideration are given.

At a later date we shall discuss the displacement
of a granular medium in the asymptotic state, i.e. same time
after the establishment of the boundary conditions, and this
condition may be regarded as being independent of time.

We shall confine our considerations to conditions
of displacement in the field of gravity.

In that case it will be convenient to adopt the
cartesian coordinate system $\{x^1, x^2, x^3\}$ with $x^3$ axis directed
vertically upward, in the opposite direction to the force of
gravity. If we confine our considerations to displacement in
the field of gravity the $x^3$ axis will be directed vertically upward. For convenience, the sequence of horizontal planes $x^3 = x_I^3 =$
$=$ const. will be designated by Roman numerals. The coordinates
of the points belonging to these planes will be designated
$x_I^1, x_I^2, x_I^3$ , or simply I. The vector field given in the plane
$x^3 = x_I^3$ defines the system of equations:

$$w^i = w^i(x_I^1, x_I^2, x_I^3) \quad \text{for} \quad i = 1, 2, 3$$

abbreviated thus

$$w^i = w^i(I) \quad \text{for} \quad i = 1, 2, 3$$

where $w^i$ are components of the vector $\bar{w}$ , or written $\bar{w} = \bar{w}(I)$.

Henceforth, the displacement of the medium will be considered in the half-space $x^3 \geqslant x_0^3$ .

We assume that the given value $w^i$ on the boundary of this region, i.e. in the plane $x^3 = x_0^3$ , defines the vector field uniquely in the whole half-space. Hence $\bar{w} = \bar{w}(I)$ uniquely defines $\bar{w} = \bar{w}(II)$ for $x^3_{II} > x^3_{I}$ which can be written symbolically as follows

$$(3.1) \qquad\qquad \bar{w}(II) = F\left[\bar{w}(I),I,II\right]$$

for any

$$x^3 = x^3_{II} > x^3_{I}$$

where $F$ denotes the uniqueness operation transforming the vector field $\bar{w}(I)$ into the vector field $\bar{w}(II)$ . This operation is dependent on $\bar{w}(I)$ and coordinates of the points of planes I and II.

If, analogously to equation 3.1 we pass from plane $x^3 = x^3_{II}$ to plane $x^3 = x^3_{III} > x^3_{I}$, we get

$$(3.2) \qquad\qquad \bar{w}(III) = F\left[\bar{w}(II),II,III\right].$$

Assuming the uniqueness of the operator $F$ , the operation of transition from $I \longrightarrow II$ , and then $II \longrightarrow III$ , must give the same effect as the transition $I \longrightarrow III$ , or

$$(3.3) \qquad\qquad \bar{w}(III) = F\left[\bar{w}(I),I,III\right].$$

Compositionof operation 3.1 and 3.2 must give the same effect as operation 3.3

$$F\left[\bar{w}(I),I,III\right] = F\left\{F\left[\bar{w}(I),I,II\right],II,III\right\} . \qquad (3.4)$$

Assumption of the uniqueness of operator $F$ leads to the relation 3.4, which is to be fulfilled for the arbitrary function
$w(I)$ . Thus, we obtain the iterated functional equation for
$F$ (of the type which occurs in Lie's group theory, or in
the theory of geometrical objects). Obviously, not every operator will fulfil the relation 3.4. However, the set of operators
fulfilling the relation is so large as to render it virtually
useless. It therefore becomes necessary to impose additional
conditions on operator $F$ , and to select an auxiliary set of
operators which fulfil those conditions.

Let the additional condition be linearity of
the operator, which is the simplest mathematical possibility.

Departure from this assumption and the attempt
to assume nonlinearity of the operator encounters difficulty
in the choice of such a nonlinearity, because the number of dif
ferent kinds of nonlinearity, that can be conceived is unlimit-
ed. Many problems can also be interpreted within a sufficiently
narrow range by means of    linear models and such models are
therefore useful from a practical standpoint.

The assumption of linearity of the operator $F$
is expressed in the form:

$$F\left[\sum_{\alpha=1}^{m} a_{\alpha}\bar{w}_{\alpha}(I),I,II\right] =$$

(3.5)

$$= \sum_{\alpha=1}^{m} a_{\alpha}F\left[\bar{w}_{\alpha}(I),I,II\right]$$

where $\bar{w}_{\alpha}$ is a sequence of $m$ vectors, and $a_{\alpha}$ is a sequence of scalar magnitudes.

The next postulate about operator $F$ is that it is a unit operator, i.e.

(3.6)                  $\lim\limits_{II\to I} F\left[\bar{w}(I),I,II\right] = \bar{w}(I)$ .

The last postulate assumes the finiteness of operator $F$ , i.e. for $\bar{w}(I) < \infty$

(3.7)                  $F\left[\bar{w}(I),I,II\right] < \infty$ .

It follows from the postulate of linearity 3.5 and from the theorem of Fréchet and Riesz, according to which every linear functional in Hilbert's space is a scalar product, that the operator $F$ is an integral operator.

This operator can be given the form:

(3.8)  $w^{k}(II) = \sum_{i=1}^{3} \int_{\Omega(I)} w^{i}(I)\varphi^{ik}(I,II)d\Omega(I)$  for  $k = 1,2,3$

where the symbol $\int...d\Omega(I)$ denotes integration of the whole area of the plane $x^{3} = x_{I}^{3}$ .

The system of the function $w^{i} = w^{i}(I)$ for $i = 1,2,3$ describes the vector field $\bar{w} = \bar{w}(I)$ in the place $x^{3} = x_{I}^{3}$ . The square matrix of function $\varphi^{ik} = \varphi^{ik}(I,II)$ is the matrix of the tran-

functions from plane $x^3 = x_I^3$ to plane $x^3 = x_{II}^3 > x_I^3$.

In accordance with postulate 3.6 operator $F$ should fulfil the condition of convergence to the boundary conditions.

For this conditions to be fulfilled, the elements of the functional matrix $\varphi^{ik}$ must fulfil the following condition

$$\lim_{II \to I} \varphi^{ik}(\underset{0}{I},II) = \varphi^{ik}(\underset{0}{I},I) = \delta^{ik}\delta(\underset{0}{I},I) \tag{3.9}$$

where

$$\delta^{ik} = \begin{cases} 0 & \text{for} \quad i \neq k \\ 1 & \text{for} \quad i = k \end{cases}$$

$$\delta(\underset{0}{I}\,I) = 0 \quad \text{for} \quad \underset{0}{I} \neq I$$

and $\int_{\Omega(I)} \delta(\underset{0}{I},I) d\Omega(I) = 1$ .

The integral operator 3.8 must fulfil the postulated semigroup condition 3.4.

If we write this condition for the integrated form of operator 3.8 in the equation 3.4, on the left side we get

$$w^l(III) = \sum_{i=1}^{} \int_{\Omega(I)} w^i(I)\varphi^{il}(I,II)d\Omega(I) \quad \text{for} \quad l = 1,2,3 \tag{3.10}$$

and the right side of the equation 3.4 will have the form

$$w^l(III) = \sum_{k=1}^{3} \int_{\Omega(II)} \sum_{i=1}^{} \int_{\Omega(I)} w^i(I)\varphi^{ik}(I,II)d\Omega(I)\varphi^{kl}(II,III)d\Omega(II) .$$

By changing the sequence of addition and integration in the last equation, we may write

$$(3.11) \quad w^\ell(\text{III}) = \sum_{i=1}^{3} \int_{\Omega(\text{I})} w^i(\text{I}) \sum_{k=1}^{3} \int_{\Omega(\text{II})} \varphi^{ik}(\text{I},\text{II}) \varphi^{k\ell}(\text{II},\text{III}) d\Omega(\text{II}) d\Omega(\text{I}) .$$

Since, according to postulate 3.4, the result of the operation $F\left[\bar{w}(\text{I}),\text{I},\text{III}\right]$ is equivalent to the result of $F\{F\left[\bar{w}(\text{I}),\text{I},\text{II}\right]\text{II},\text{III}\}$ , and the relations 3.10 and 3.11 are to be fulfilled for the arbitrary vector fields $w^i = w^i(\text{I})$, then

$$(3.12) \quad \varphi^{i\ell}(\text{I},\text{III}) = \sum_{k=1}^{3} \int_{\Omega(\text{II})} \varphi^{ik}(\text{I},\text{II}) \varphi^{k\ell}(\text{II},\text{III}) d\Omega(\text{II}) \qquad \text{for} \quad i,\ell = 1,2,3$$

must be fulfilled.

The system of integral equations 3.12 is a generalization of the so-called equation of Smoluchowski describing the class of Markow's continuous basic processes in the theory of diffusion and Brownian movements.

The system of equations 3.12 is the basic equation for interpretation of displacement of granular media from standpoint of the concept presented herein.

One class of solutions of the system 3.12 can be obtained by means of the integral transformation of Fourier, and another by constructing a system of differential equations similar to the method employed by Kolmogorow.

# 4. METHOD OF SOLUTION OF THE SYSTEM BY MEANS OF FOURIER'S TRANSFORMATION [5,4]

This method can be employed in the special case when the transitional functions $\varphi^{ik}$ are homogeneous functions (physically, this corresponds to a homogeneous medium), i.e. if these functions depend only on the difference of coordinates.

In that case

$$\varphi^{ik}(I,II) \;=\; \varphi^{ik}(II-I) \;=\; \varphi^{ik}\big(x_{II}^1 - x_I^1, x_{II}^2 - x_I^2, x_{II}^3 - x_I^3\big). \qquad (4.1)$$

As assumed, these functions belong to the class of integrable functions (according to Lebesgue), and their Fourier trasformants exist, which will be designated by capital letters

$$\phi^{ik}\big(\xi,\eta,x_{II}^3 - x_I^3\big) \;=$$

$$= \int\limits_{-\infty}^{+\infty}\int\limits_{-\infty}^{+\infty} \varphi^{ik}\big(x_{II}^1 - x_I^1, x_{II}^2 - x_I^2, x_{II}^3 - x_I^3\big) e^{i[\xi(x_{II}^1 - x_I^1) + \eta(x_{II}^2 - x_I^2)]}\, dx_{II}^1\, dx_{II}^2 . \qquad (4.2)$$

The reverse transformation has the form:

$$\varphi^{ik}\big(x_{II}^1 - x_I^1, x_{II}^2 - x_I^2, x_{II}^3 - x_I^3\big) \;=$$

$$= \frac{1}{4\pi^2} \int\limits_{-\infty}^{+\infty}\int\limits_{-\infty}^{+\infty} \phi^{ik}\big(\xi,\eta,x_{II}^3 - x_I^3\big) e^{-i[\xi(x_{II}^1 - x_I^1) + \eta(x_{II}^2 - x_I^1)]}\, d\xi\, d\eta . \qquad (4.3)$$

Fourier's transformation will be applied to the system 3.12. To this end, new variables are introduced

$$x^1_{III} - x^1_I = \alpha; \quad x^2_{III} - x^2_I = \delta; \quad x^3_{III} - x^3_I = \varrho$$

$$(4.4) \quad x^1_{III} - x^1_{II} = \beta; \quad x^2_{III} - x^2_{II} = \varepsilon; \quad x^3_{III} - x^3_{II} = \vartheta$$

$$x^1_{II} - x^1_I = \gamma; \quad x^2_{II} - x^2_I = \zeta; \quad x^3_{II} - x^3_I = \tau.$$

After introduction of these variables, equation 3.12 assumes the form

$$(4.5) \quad \varphi^{il}(\alpha\,\delta\,\varrho) = \sum_{k=1}^{3} \int_{-\infty}^{+\infty} \int_{-\infty}^{+\infty} \varphi^{ik}(\gamma,\zeta,\tau)\varphi^{kl}(\alpha - \gamma, \delta - \zeta, \vartheta)\,d\gamma\,d\zeta .$$

The last equation is submitted to Fourier's transformation, giving the system of functional equations

$$(4.6) \quad \phi^{il}(\xi,\eta,\tau+\vartheta) = \sum_{k=1}^{3} \phi^{ik}(\xi,\eta,\tau)\phi^{kl}(\xi,\eta,\vartheta) \qquad \text{for} \quad i,l = 1,2,3 .$$

In order to show that equation 4.6 is a result of Fourier transformation, it must be demonstrated that the application of reverse Fourier transformation to the right side of 4.6 gives the right side of equation 4.5. Taking into account 4.2 we get

$$\frac{1}{4\pi^2} \sum_{k=1}^{3} \int_{-\infty}^{+\infty} \int_{-\infty}^{+\infty} \phi^{ik}(\xi,\eta,\tau)\phi^{kl}(\xi,\eta,\vartheta) e^{-i(\xi\alpha+\eta\delta)}\,d\xi\,d\eta =$$

$$= \frac{1}{4\pi^2} \sum \int_{-\infty}^{+\infty} \int_{-\infty}^{+\infty} \phi^{ki}(\xi,\eta,\vartheta) \int_{-\infty}^{+\infty} \int_{-\infty}^{+\infty} \varphi^{ik}(\gamma,\zeta,\tau) e^{i(\xi\gamma+\eta\zeta)}\,d\gamma\,d\zeta\, e^{-i(\xi\alpha+\eta\delta)}\,d\xi\,d\eta .$$

Changing the sequence of integration in the last expression and taking into account 4.3 we get

$$\sum_{R=1}^{3} \int_{-\infty}^{+\infty} \int_{-\infty}^{+\infty} \varphi^{ik}(\gamma,\zeta,\tau) \left\{ \frac{1}{4\pi^2} \int_{-\infty}^{+\infty}\int_{-\infty}^{+\infty} \phi^{kl}(\xi,\eta,\vartheta) e^{-i[\xi(\alpha-\gamma)+\eta(\vartheta-\zeta)]} d\xi\, d\eta \right\} d\gamma\, d\zeta =$$

$$= \sum_{k=1}^{3} \int_{-\infty}^{+\infty}\int_{-\infty}^{+\infty} \varphi^{ik}(\gamma,\zeta,\tau)\, \varphi^{kl}(\alpha-\gamma,\vartheta-\zeta,\vartheta)\, d\gamma\, d\zeta \; .$$

(4.7)

The reverse transformant of the right side of equation 4.6 has the form of equation 4.7; and the transformant of the last expression has the form of the right side of the last expression 4.6.

In these equations, the variables $\xi$ and $\eta$ are parameters, and the variables $\vartheta, \tau, \vartheta+\tau$ are arguments.

Solutions $\phi^{ik}$ of system 4.6 by application of Fourier transformation 4.3 gives the solutions $\varphi^{ik}$ of the system of integral equations 3.12.

The solutions $\varphi^{ik}$ fulfil the boundary condition 3.9. In order to obtain a suitable boundary condition for function $\phi^{ik}$ , Fourier's transformation must be applied to the Dirac "function" on the right side of equation 3.9. This gives

$$\int_{-\infty}^{+\infty}\int_{-\infty}^{+\infty} \delta^{ik}\, \delta(I,I)\, e^{i[\xi(x_I^1-x^1)+\eta(x_I^2-x^2)]} df(I) = \delta^{ik}$$

where $\delta^{ik}$ is Kronecker's symbol.

According to the last equation, we seek the solution of the system of the functional equation 4.6 in order to fulfil the boundary condition

$$\Phi^{ik}(\xi,\eta,0) = \delta^{ik} .$$

The system of functional equations 4.6 has been studied by a number of authors such as Ghermanescu [6] and by Gołab and Kraj.

The last mentioned authors reduce the problem to the solution of a system of differential equations. Since in equation 4.6 $\xi$ and $\eta$ are parameters, and $\tau, \vartheta$ and $\vartheta+\tau$ are arguments, the system can be written more clearly as follows

$$(4.8) \qquad \Phi^{il}(\vartheta + \tau) = \sum_{k=1}^{3} \Phi^{ik}(\tau)\Phi^{kl}(\vartheta)$$

for boundary conditions

$$(4.9) \qquad \Phi^{ik}(0) = \delta^{ik} .$$

In addition, we assume that the derivative values are known

$$(4.10) \qquad \left[\frac{d\Phi^{kl}(s)}{ds}\right]_{s=0} = \Lambda^{kl} .$$

Differentiating the system of equations 4.8 in relation to $\vartheta$ and assuming $\vartheta = 0$, we get

$$(4.11) \qquad \frac{d\Phi^{il}(\tau)}{d\tau} = \sum_{k=1}^{3} \Lambda^{kl}\Phi^{ik}(\tau) \quad \text{for} \quad i,l = 1,2,3 .$$

This is a system of linear differential equations of the first order, in which the coefficients $\Lambda^{kl}$ are dependent on the parameters $\xi$ and $\eta$ .

The characteristic equation of this system has

the form

$$Det(\Lambda^{kl} - \sigma\delta^{kl}) = 0.$$

Let the equation have different characteristic roots $\overset{1}{\sigma}, \overset{2}{\sigma}, \overset{3}{\sigma}$

The general solution of system 4.11 has the form

$$\phi^{il}(\tau) = \sum_{\alpha=1}^{3} \overset{il}{\underset{\alpha}{A}} \exp(\overset{\alpha}{\sigma}\tau) \tag{4.12}$$

where the coefficients $\overset{il}{\underset{\alpha}{A}}$ depend on the parameters $\xi$ and $\eta$ .

These coefficients should be selected to fulfil

the boundary conditions 4.9 and 4.10.

According to equation 4.9

$$\sum_{\alpha=1}^{3} \overset{il}{\underset{\alpha}{A}} = \delta^{il} \tag{4.13}$$

must be fulfilled, and according to 4.10

$$\sum_{\alpha=1}^{3} \overset{\alpha}{\sigma} \overset{ik}{A} = \Lambda^{ik}. \tag{4.14}$$

Hence, in order to satisfy the initial condition

4.9 and 4.10, the coefficients $\overset{ik}{\underset{\alpha}{A}}$ must satisfy the conditions

4.13 and 4.14.

Having the solution 4.12, we get the solution of

4.3 in the form

$$\varphi^{il}(I,II) = \frac{1}{4\pi^2} \int_{-\infty}^{+\infty} \int_{-\infty}^{+\infty} \sum_{\alpha=1}^{3} \overset{il}{\underset{\alpha}{A}}(\xi,\eta) \exp\left[\overset{\alpha}{\sigma}(\xi,\eta)(x^3_{II}-x^3_I)\right] - i\left[\xi(x^1_{II}-x^1_I)+\eta(x^2_{II}-x^2_I)\right] d\xi d\eta.$$

Having the matrix of the solutions $\varphi^{il}(I,II)$, and employing the operator 3.8, we may solve the boundary problem consisting in determination of the vector field $\bar{w} = \bar{w}(\mathrm{II})$ if the vector field $\bar{w} = \bar{w}(\mathrm{I})$ is given.

Example. As a special example, let us consider a two dimensional plane state of displacement of the medium.

For this purpose, we may take the Cartesian system of coordinates $\{x, z\}$ with the $z$ axis directed vertically upward. Such a medium is homogeneous, i.e. the functions of the transformation $\varphi^{ij}$ are homogeneous, and consequently dependent only on the difference of the coordinates $x_{\mathrm{II}} - x_{\mathrm{I}}$ and $z_{\mathrm{II}} - z_{\mathrm{I}}$ .

In the case of a two-dimensional plane state of displacement, the matrix of the transition functions has the form

$$\begin{pmatrix} \varphi^{11}, & \varphi^{12} \\ \varphi^{21}, & \varphi^{22} \end{pmatrix} .$$

The system of equations 3.12 now takes the form

$$\varphi^{il}(x_{\mathrm{III}} - x_{\mathrm{I}}, z_{\mathrm{III}} - z_{\mathrm{I}}) = \sum_{k=1}^{2} \int_{-\infty}^{+\infty} \varphi^{ik}(x_{\mathrm{II}} - x_{\mathrm{I}}, z_{\mathrm{II}} - z_{\mathrm{I}})\, \varphi^{kl}(x_{\mathrm{III}} - x_{\mathrm{II}}, z_{\mathrm{III}} - z_{\mathrm{II}})\, dx_{\mathrm{II}}$$

for $i, l = 1, 2$ .

Substituting

$$x_{\mathrm{III}} - x_{\mathrm{I}} = \alpha \qquad z_{\mathrm{III}} - z_{\mathrm{I}} = t + \tau$$

$$x_{\mathrm{II}} - x_{\mathrm{I}} = \beta \qquad z_{\mathrm{II}} - z_{\mathrm{I}} = t$$

we get

$$\varphi^{il}(\alpha, t + \tau) = \sum_{k=1}^{2} \int_{-\infty}^{+\infty} \varphi^{ik}(\beta, t)\, \varphi^{kl}(\alpha - \beta, \tau)\, d\beta$$

and using Fourier's transformation for the last system of equa-

tions we get

$$\int_{-\infty}^{+\infty} e^{i\xi}\varphi^{il}(\alpha,t+\tau)d\alpha =$$

$$= \sum_{k=1}^{2}\int_{-\infty}^{+\infty}\int_{-\infty}^{+\infty} e^{i\xi\alpha}\varphi^{ik}(\beta,t)\varphi^{kl}(\alpha-\beta,\tau)d\beta d\alpha .$$

(4.15)

Substituting: $\alpha-\beta=m, \beta=n$ on the right side of the equation

4.15 we get

$$\sum_{k=1}^{2}\int_{-\infty}^{+\infty}\int_{+\infty}^{+\infty} e^{i\xi(m+n)}\varphi^{ik}(n,t)\varphi^{kl}(m,\tau)dmdn =$$

$$= \sum_{k=1}^{2}\int_{-\infty}^{+\infty} e^{i\xi n}\varphi^{ik}(n,t)dn\int_{-\infty}^{+\infty} e^{i\xi m}\varphi^{kl}(m,\tau)dm .$$

(4.16)

By determining the transformates

$$\int_{-\infty}^{+\infty} e^{i\xi\alpha}\varphi^{kl}(\alpha,s)d\alpha = \phi^{kl}(\xi s)$$

(4.17)

we get from the equations 4.15 and 4.16

$$\phi^{kl}(\xi,t+\tau) = \sum_{h=1}^{2}\phi^{kn}(\xi,t)\phi^{nl}(\xi,\tau)$$

(4.18)

$$\text{for } k,l = 1,2$$

where $\xi$ is a parameter, and $t,\tau$ are arguments. The solution of

the system of functional equations 4.18 may be reduced to the so

lutions of the systems of ordinary differential equations 4.11

taking into account the initial conditions 4.9 and 4.10, i.e.

$$\phi^{ik}(0) = \delta^{ik}$$

(4.19)

$$\left(\frac{d\phi^{k\ell}(s)}{ds}\right)_{s=0} = \Lambda^{k\ell}.$$

In the case under consideration, the system of differential e-
quations 4.11 takes the form

(4.20)
$$\frac{d\phi^{i\ell}(\tau)}{d\tau} = \sum_{k=1}^{2} \Lambda^{k\ell} \phi^{ik}(\tau)$$

where $\Lambda^{k\ell}$ in general depends on the parameter $\xi$ .

The solution of the last system of equations for
the initial conditions 4.19 has the form

$$\phi^{11}(\xi,\tau) = \varrho(e^{\sigma_1\tau} - e^{\sigma_2\tau}) + e^{\sigma_2\tau}$$

$$\phi^{12}(\xi,\tau) = \eta(e^{\sigma_1\tau} - e^{\sigma_2\tau})$$

$$\phi^{21}(\xi,\tau) = \zeta(e^{\sigma_1\tau} - e^{\sigma_2\tau})$$

$$\phi^{22}(\xi,\tau) = \varrho(e^{\sigma_2\tau} - e^{\sigma_1\tau}) + e^{\sigma_1\tau}$$

and

$$\sigma_{1,2} = \frac{1}{2}\left\{\Lambda^{11} + \Lambda^{22} \pm \left[(\Lambda^{11} - \Lambda^{22})^2 + 4\Lambda^{12}\Lambda^{21}\right]^{\frac{1}{2}}\right\}$$

and

(4.21) $\eta = \dfrac{\Lambda^{12}}{\sigma_1 - \sigma_2}$ ; $\quad \zeta = \dfrac{\Lambda^{21}}{\sigma_1 - \sigma_2}$ ; $\quad \varrho = \dfrac{1}{2} + \left(\dfrac{1}{4} - \eta\zeta\right)^{\frac{1}{2}}$ .

For the solution of $\psi^{ik}$ , the transformation
4.17 must be reversed.

This can be simplified if certain properties of the transitional function $\varphi^{ik}$ following from the subsequent considerations are taken into account.

For instance, let us take the relation 3.8 applied to the two-dimensional plane problem of displacement.

Writing out the relation in this particular case gives

$$w^1(II) = \int_{-\infty}^{+\infty}\left[w^1(I)\varphi^{11}(I,II) + w^2(I)\varphi^{21}(I,II)\right]dx_1$$
$$w^2(II) = \int_{-\infty}^{+\infty}\left[w^1(I)\varphi^{12}(I,II) + w^2(I)\varphi^{22}(I,II)\right]dx_1 . \tag{4.22}$$

We assume that the initial conditions for $x^3 = x^3_I$ at level I are given for the component displacements in the form of so-called Dirac "function". For instance

$$w^1(I) = \delta(I - \underset{0}{I}), \qquad w^2(I) = 0 .$$

From relation 4.22 we get

$$w^1(II) = \varphi^{11}(\underset{0}{I}, II)$$
$$w^2(II) = \varphi^{12}(\underset{0}{I}, II) .$$

If the medium in which the field of displacements is accomplished is a horizontally homogeneous medium, it will be natural to assume that function $\varphi^{11}$ with regard to the $x$ coordinates (plotted on the horizontal axis) is an even function, i.e.

(4.23)        $\varphi^{11}(x_2 - x_1, z_2 - z_1) = \varphi^{11}(x_1 - x_2, z_2 - z_1)$ .

Analogously, for a horizontally homogeneous medium we get

(4.24)        $\varphi^{12}(x_2 - x_1, z_2 - z_1) = -\varphi^{12}(x_1 - x_2, z_2 - z_1)$

i.e. function $\varphi^{12}$ with respect to coordinate $x$ is an odd function.

       Let us now take the initial conditions in the

form

(4.25)            $w^1(I) = 0, \quad w^2(I) = \delta(I - \underset{0}{I})$

from the relation 4.22 we get

$$w^1(II) = \varphi^{21}(\underset{0}{I}, II)$$

$$w^2(II) = \varphi^{22}(\underset{0}{I}, II) .$$

Satisfying the condition 4.25 in a horizontally homogeneous medium we get

(4.26)        $\varphi^{21}(x_2 - x_1, z_2 - z_1) = -\varphi^{21}(x_1 - x_2, z_2 - z_1)$

or, the transition function $\varphi^{21}$ with respect to the variable $x$
is an odd function.

       According

(4.27)        $\varphi^{22}(x_2 - x_1, z_2 - z_1) = \varphi^{22}(x_1 - x_2, z_2 - z_1)$

with respect to the variable $x$ is an even function.

       As a result of the evenness and oddness of the

function $\varphi^{kl}$ defined by the equations 4.23, 4.24, 4.26 and 4.27, reverse transformation to transformation 4.17 takes the form

$$\varphi^{kl}(x,\tau) = \frac{1}{\pi} \int_0^\infty \phi^{kl}(\xi,\tau)\cos(\xi x)d\zeta \qquad \text{for} \qquad k = l$$

and                                                                                   (4.28)

$$\varphi^{kl}(x,\tau) = \frac{1}{\pi} \int_0^\infty \phi^{kl}(\xi,\tau)\sin(\xi x)d\zeta \qquad \text{for} \qquad k \neq l.$$

As a result of the oddness of functions $\varphi^{12}$ and $\varphi^{21}$

$$\int_{-\infty}^{+\infty} \varphi^{lk}(x_2 - x_1, z_2 - z_1)dx_2 = 0 \qquad (4.29)$$

is satisfied.

By reversing the functions $\phi^{kl}$ in agreement with equations 4.28 and having the transition functions $\varphi^{kl}$, we can solve the boundary value problem by means of the equation 3.8.
Applying transfor<u>r</u> mation 4.28 and a<u>s</u> suming certain

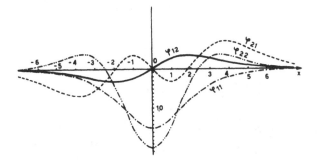

Fig. 12

special forms of coefficients $\Lambda^{kl} = \Lambda^{kl}(\xi)$ in equation 4.20, in effect we get the solution shown in Fig. 12.

## 5. THE SECOND METHOD OF SOLVING THE SYSTEM OF SMOLUCHOWSKI EQUATIONS (3.12) [1,4]

The idea of this method is similar to the one used by Kolmogorow in his work on stochastic processes. Compared with the latter, it is a generalization with real physical sense.

In this method, certain conditions is imposed on the class of function $\varphi^{ik}$ . The conditions concern the moments of the transition function.

Namely, we assume the limits

(5.1)
$$\lim_{\substack{x^3_{III} \to x^3_{II}}} \frac{1}{x^3_{III} - x^3_{II}} \int_{-\infty}^{+\infty}\int_{-\infty}^{+\infty} \varphi^{ik}(II,III)(x^\alpha_{III} - x^\alpha_{II})df(III) = A^{ik}_\alpha(II)$$
$$\text{for } \alpha = 1,2$$

(5.2)
$$\lim_{\substack{x^3_{III} \to x^3_{II}}} \frac{1}{x^3_{III} - x^3_{II}} \int_{-\infty}^{+\infty}\int_{-\infty}^{+\infty} \varphi^{ik}(II,III)(x^\alpha_{III} - x^\alpha_{II})(x^\beta_{III} - x^\beta_{II})df(III) = B^{ik}_{\alpha\beta}(II)$$
$$\text{for } \alpha,\beta = 1,2$$

(5.3)
$$\lim_{\substack{x^3_{III} \to x^3_{II}}} \frac{1}{x^3_{III} - x^3_{II}} \int_{-\infty}^{+\infty}\int_{-\infty}^{+\infty} \varphi^{ik}(II,III)(x^\alpha_{III} - x^\alpha_{II})(x^\beta_{III} - x^\beta_{II})(x^\gamma_{III} - x^\gamma_{II})df(III) = 0$$
$$\text{for } \alpha,\beta,\gamma = 1,2$$

and

(5.4)
$$\int_{-\infty}^{+\infty}\int_{-\infty}^{+\infty} \varphi^{ik}(II,III)df(III) = \delta^{ik} + N^{ik}(II)(x^3_{III} - x^3_{II}).$$

The first assumptions 5.1, 5.2 and 5.3 indicate that for $x^3_{III} \rightarrow x^3_{II}$ the transition functions $\varphi^{ik}$ are "concentrated" in point II, that everywhere beyond point II the value of function $\varphi^{ik}$ tends towards zero, and that $\varphi^{ik}$ have finite moments of the first, second and third order.

Physically, this may be interpreted as showing that the influence of the initial conditions in point II, e.g. in the form of Dirac's function, as $x^3_{III} \rightarrow x^3_{II}$ disappears completely beyond the cylinder with its base in plane II and centre in point II, and height $(x^3_{III} - x^3_{II})$.

The assumption 5.4 follows from the suggestion given in the example on pages 50–55 and the properties of evenness and oddness of functions $\varphi^{ik}$ 4.23, 4.24, 4.26, and 4.27 and its physical interpretation.

A consequence of these properties is equation 4.29 relating to the added transition functions $\varphi^{ik}$. Interpretation of the expression $N^{ik}$ on the right side of equation 5.4 is connected with the change in volume of the "trough" formed in the granular medium.

Below we give an outline of the method of introducing the system of differential equations whose integrals satisfy Smoluchowski's system of integral equations 3.12.

In order to introduce this system, we introduce the following expression:

$$
(5.5) \qquad \mathfrak{z}^{ik} = \int_a^b \int_c^d \frac{\partial}{\partial x^3_{II}} \varphi^{ik}(I, x^1_{III}, x^2_{III}, x^3_{II}) R(x^1_{III}, x^2_{III}) df(III)
$$

in which $R = R(x^1_{III}, x^2_{III})$ is a function which

1) in the rectangle **a b c d** is positive

2) in the rectangle **a b c d** is of class $C^3$

3) on the boundary of **a b c d** is equal to zero with the
   first derivatives

4) outside of rectangle **a b c d** is equal to zero.

The expression 5.5 can be transformed by substituting the derivative according to $x^3_{II}$ by the limit of the difference quotient

$$
\mathfrak{z}^{ik} = \lim_{x^3_{III} \to x^3_{II}} \frac{1}{x^3_{III} - x^3_{II}} \int_{-\infty}^{+\infty} \int_{-\infty}^{+\infty} R(x^1_{III}, x^2_{III})\left[\varphi^{ik}(I; III) - \varphi^{ik}(I; x^1_{III}, x^2_{III}, x^3_{II})\right] df(III) .
$$

Taking into account equation 3.12 and substituting in the last equation $\varphi^{ik} = \varphi^{ik}(I, III)$ we get

$$
\mathfrak{z}^{ik} = \lim_{x^3_{III} \to x^3_{II}} \frac{1}{x^3_{III} - x^3_{II}} \int_{-\infty}^{+\infty} \int_{-\infty}^{+\infty} R(x^1_{III}, x^2_{III})\left[\sum_{n=1}^{3} \int_{-\infty}^{+\infty} \int_{-\infty}^{+\infty} \varphi^{in}(I; II)\varphi^{nk}(II; III) df(II) - \varphi^{ik}(I; x^1_{III}, x^2_{III}, x^3_{II})\right] df(III).
$$

By developing the function $R = R(x^1_{III}, x^2_{III})$ in Taylor's series in the first member of the integration of the right side of the last equation around the point $x^1_{II}$, $x^2_{II}$, we get

$$
\mathfrak{z}^{ik} = \lim_{x^3_{III} \to x^3_{II}} \frac{1}{x^3_{III} - x^3_{II}} \int_{-\infty}^{+\infty} \int_{-\infty}^{+\infty} \Bigg\{\Big[R(II) + \sum_{\alpha=1}^{2} \frac{\partial R}{\partial x^\alpha_{III}}(x^\alpha_{III} - x^\alpha_{II}) + \sum_{\alpha=1}^{2}\sum_{\beta=1}^{2} \frac{\partial^2 R}{\partial x^\alpha_{II} \partial x^\beta_{II}}.
$$

$$
(5.6a) \qquad (x^\alpha_{III} - x^\alpha_{II})(x^\beta_{III} - x^\beta_{II}) + \dots\Big] \sum_{n=1}^{3} \int_{-\infty}^{+\infty} \int_{-\infty}^{+\infty} \varphi^{in}(I, II)\, \varphi^{nk}(II, III) d f(II) -
$$

$$- R(III)\varphi^{ik}(I;x^1_{III},x^2_{III},x^3_{II})\Big\} d\,f\,(III)\,. \tag{5.6b}$$

Considered in the last integrals the expressions multiplying by functions $R(II)$ and $R(III)$ we have

$$\lim_{\substack{x^3_{III}\to x^3_{II}\\ x^3_{III}-x^3_{II}}}\frac{1}{x^3_{III}-x^3_{II}}\int_{-\infty}^{+\infty}\int_{-\infty}^{+\infty}R(II)\sum_{n=1}^{3}\varphi^{in}(I;II)\varphi^{nk}(II,III)d\,f\,(II)-R(III)\varphi^{ik}(I;x^1_{III},x^2_{III},x^3_{II})d\,f\,(III)\,.$$

Changing the sequence of integration gives

$$\lim_{\substack{x^3_{III}\to x^3_{II}\\ x^3_{III}-x^3_{II}}}\frac{1}{x^3_{III}-x^3_{II}}\Bigg[\int_{-\infty}^{+\infty}\int_{-\infty}^{+\infty}R(II)\sum_{n=1}^{3}\varphi^{in}(I;II)\int_{-\infty}^{+\infty}\int_{-\infty}^{+\infty}\varphi^{nk}(II,III)d\,f\,(III)\,d\,f\,(II)\,+$$

$$-\int_{-\infty}^{+\infty}\int_{-\infty}^{+\infty}R(II)\varphi^{ik}(I;x^1_{II},x^2_{II},x_{II})d\,f\,(II)\Bigg]\,.$$

Taking into account equation 5.4, we get:

$$\lim_{\substack{x^3_{III}\to x^3_{II}\\ x^3_{III}-x^3_{II}}}\frac{1}{x^3_{III}-x^3_{II}}\Bigg[\int_{-\infty}^{+\infty}\int_{-\infty}^{+\infty}R(II)\sum_{n=1}^{3}\varphi^{in}(I,II)\big[\delta^{nk}+N(II)^{nk}(x^3_{III}-x^3_{II})\big]d\,f\,(II)\,+$$

$$-\int_{-\infty}^{+\infty}\int_{-\infty}^{+\infty}R(II)\varphi^{ik}(I,II)d\,f\,(II)\Bigg]=\int_{-\infty}^{+\infty}\int_{-\infty}^{+\infty}R(II)\sum_{n=1}^{3}\varphi^{in}(I,II)N^{nk}(II)d\,f\,(II)\,. \tag{5.7}$$

Let us now consider the expressions of the deriv atives, $\dfrac{\partial R}{\partial x^{\alpha}_{II}}$ under the integral on the right side of the equation 5.6. Changing the sequence of integration and summation

$$\lim_{\substack{x^3_{III}\to x^3_{II}\\ x^3_{III}-x^3_{II}}}\frac{1}{x^3_{III}-x^3_{II}}\int_{-\infty}^{+\infty}\int_{-\infty}^{+\infty}\sum_{\alpha=1}^{2}\frac{\partial R}{\partial x^{\alpha}_{II}}(x^{\alpha}_{III}-x^{\alpha}_{II})\sum_{n=1}^{3}\int_{-\infty}^{+\infty}\int_{-\infty}^{+\infty}\varphi^{in}(I;II)\varphi^{nk}(II;III)d\,f(II)d\,f(III)=$$

$$=\lim_{\substack{x^3_{III}\to x^3_{II}\\ x^3_{III}-x^3_{II}}}\frac{1}{x^3_{III}-x^3_{II}}\int_{-\infty}^{+\infty}\int_{-\infty}^{+\infty}\sum_{n=1}^{3}\varphi^{in}(I;II)\sum_{\alpha=1}^{2}\int_{-\infty}^{+\infty}\int_{-\infty}^{+\infty}\frac{\partial R(II)}{\partial x^{\alpha}_{II}}\varphi^{nk}(II;III)(x^{\alpha}_{III}-x^{\alpha}_{II})d\,f\,(III)d\,f\,(II)\,.$$

According to the assumption 5.1, the right side
of the last equation may be written:

$$\sum_{n=1}^{3} \sum_{\alpha=1}^{2} \int_{-\infty}^{+\infty} \int_{-\infty}^{+\infty} \varphi^{in}(I;II) \frac{\partial R(II)}{\partial x_{II}^{\alpha}} A_{\alpha}^{nk}(II) d f(II) .$$

Integration by parts in the last expression, tak
ing into account the assumption of function $R = R(II)$, and its
derivatives, gives

(5.8)
$$-\int_{-\infty}^{+\infty} \int_{-\infty}^{+\infty} \sum_{n=1}^{3} \sum_{\alpha=1}^{2} R(II) \frac{\partial}{\partial x_{II}^{\alpha}} \Big[ \varphi^{in}(I;II) A_{\alpha}^{nk}(II) \Big] d f(II)$$

The third expression of the derivative $\dfrac{\partial^2 R}{\partial x_{II}^{\alpha} \partial x_{II}^{\beta}}$ on the right side
of equation 5.6 may now be written.

We change the sequence of integration and summation thus

$$\lim_{\substack{x_{III}^3 \to x_{II}^3 \\ }} \frac{1}{x_{II}^3 - x_{III}^3} \int_{-\infty}^{+\infty} \int_{-\infty}^{+\infty} \sum_{\alpha=1}^{2} \sum_{\beta=1}^{2} \frac{\partial^2 R}{\partial x_{II}^{\alpha} \partial x_{II}^{\beta}} (x_{III}^{\alpha} - x_{II}^{\alpha})(x_{III}^{\beta} - x_{II}^{\beta}) \sum_{n=1}^{3} \int_{-\infty}^{+\infty} \int_{-\infty}^{+\infty} \varphi^{in}(I,II) \varphi^{nk}(II,III) d f(II) d f(III) =$$

$$= \lim_{\substack{x_{III}^3 \to x_{II}^3 \\ }} \frac{1}{x_{II}^3 - x_{III}^3} \int_{-\infty}^{+\infty} \int_{-\infty}^{+\infty} \sum_{n=1}^{3} \varphi^{in}(I;II) \sum_{\alpha=1}^{2} \sum_{\beta=1}^{2} \int_{-\infty}^{+\infty} \int_{-\infty}^{+\infty} \frac{\partial^2 R}{\partial x_{II}^{\alpha} \partial x_{II}^{\beta}} \varphi^{nk}(II,III)(x_{III}^{\alpha} - x_{II}^{\alpha})(x_{III}^{\beta} - x_{II}^{\beta}) d f(III) d f(II) .$$

Taking into account equation 5.2, the right side
of the last equation may be written

$$\sum_{n=1}^{3} \sum_{\alpha=1}^{2} \sum_{\beta=1}^{2} \int_{-\infty}^{+\infty} \int_{-\infty}^{+\infty} \varphi^{in}(I;II) \frac{\partial^2 R(II)}{\partial x_{II}^{\alpha} \partial x_{II}^{\beta}} B_{\alpha\beta}^{nk}(II) d f(II) .$$

Integrating twice by parts in the last expression

and taking into account the assumption of function $R = R(II)$ and its derivatives, gives

$$\int_{-\infty}^{+\infty}\int_{-\infty}^{+\infty} \sum_{n=1}^{3}\sum_{\alpha=1}^{2}\sum_{\beta=1}^{2} R(II) \frac{\partial^2}{\partial x^{\alpha}_{II}\partial x^{\beta}_{II}}\left[\varphi^{in}(I,II)B^{nk}_{\alpha\beta}(II)\right]d\,f\,(II)\,. \qquad (5.9)$$

The introduction of the calculations of expres-sions 5.7, 5.8 and 5.9 into equation 5.6 gives

$$J^{ik} = \int_{-\infty}^{+\infty}\int_{-\infty}^{+\infty}\left\{\sum_{n=1}^{3}\sum_{\alpha=1}^{2}\sum_{\beta=1}^{2}\frac{\partial^2}{\partial x^{\alpha}_{II}\partial x^{\beta}_{II}}\left[\varphi^{in}(I;II)B^{nk}_{\alpha\beta}(II)\right] + \right.$$

$$\left. - \sum_{n=1}^{3}\sum_{\alpha=1}^{2}\frac{\partial}{\partial x^{\alpha}_{II}}\left[\varphi^{in}(I;II)A^{nk}_{\alpha}(II)\right] + \sum_{n=1}^{3}\varphi^{in}(I;II)N^{nk}(II)\right\}R(x^1_{II},x^2_{II})df(II)\,.$$

On comparing the latter equation with equation 5.5 and taking into account that $R=R(x^1,x^2)$ is an arbitrary func-tion, we get

$$\frac{\partial\varphi^{ik}(I;II)}{\partial x^3_{II}} = \sum_{n=1}^{3}\sum_{\alpha=1}^{2}\sum_{\beta=1}^{2}\frac{\partial^2}{\partial x^{\alpha}_{II}\partial x^{\beta}_{II}}\left[\varphi^{in}(I;II)B^{nk}_{\alpha\beta}(II)\right] +$$

$$- \sum_{n=1}^{3}\sum_{\alpha=1}^{2}\frac{\partial}{\partial x^{\alpha}_{II}}\left[\varphi^{in}(I;II)A^{nk}_{\alpha}(II)\right] + \sum_{n=1}^{3}\varphi^{in}(I;II)\,N^{nk}(II)$$

$$\text{for } i,k =1,2,3\,. \qquad (5.10)$$

From the last system of equations 5.10 we may obtain the system of differential equations defining the vector field $w^k$.

To this end, we multiply the equations of the sys-tem 5.10 by $w^i = w^i(I)$, and then integrate according to $x^1_I, x^2_I$ and add up according to $i$, obtaining

$$\sum_{i=1}^{3} \int\limits_{-\infty}^{+\infty}\int\limits_{-\infty}^{+\infty} w^{i}(I)\frac{\partial \varphi^{ik}(I;II)}{\partial x^{3}}\,df(I) =$$

$$= \sum_{n=1}^{3}\sum_{\alpha=1}^{2}\sum_{\beta=1}^{2} \int\limits_{-\infty}^{+\infty}\int\limits_{-\infty}^{+\infty}\sum_{i=1}^{3} w^{i}(I)\frac{\partial^{2}}{\partial x^{\alpha}\partial x^{\beta}}_{II\ II}[\varphi^{in}(I;II)B^{nk}_{\alpha\beta}(II)]\,df(I) +$$

$$-\sum_{n=1}^{3}\sum_{\alpha=1}^{2} \int\limits_{-\infty}^{+\infty}\int\limits_{-\infty}^{+\infty}\sum_{i=1}^{3} w^{i}(I)\frac{\partial}{\partial x^{\alpha}}_{II}[\varphi^{in}(I;II)A^{nk}_{\alpha}(II)]\,df(I) + \sum_{n=1}^{3} \int\limits_{-\infty}^{+\infty}\int\limits_{-\infty}^{+\infty}\sum_{i=1}^{3} w^{i}(I)\varphi^{in}(I;II)N^{nk}(II)\,df(I).$$

by changing the order of integration and summation we get

$$\frac{\partial w^{k}(II)}{\partial x^{3}}_{II} = \sum_{n=1}^{3}\sum_{\alpha=1}^{2}\sum_{\beta=1}^{2} \frac{\partial^{2}}{\partial x^{2}\partial x^{\beta}}_{II\ II}\left[\left\{\sum_{i=1}^{3} \int\limits_{-\infty}^{+\infty}\int\limits_{-\infty}^{+\infty} w^{i}(I)\varphi^{in}(I;II)\,df(I)\right\}B^{nk}_{\alpha\beta}(II)\right] +$$

$$-\sum_{n=1}^{3}\sum_{\alpha=1}^{2}\frac{\partial}{\partial x^{\alpha}}_{II}\left[\left\{\sum_{i=1}^{3}\int\limits_{-\infty}^{+\infty}\int\limits_{-\infty}^{+\infty} w^{i}(I)\varphi^{in}(I;II)\,df(I)\right\}A^{nk}_{\alpha}(II)\right] + \sum_{n=1}^{3}\left\{\sum_{i=1}^{3}\int\limits_{-\infty}^{+\infty}\int\limits_{-\infty}^{+\infty} w^{i}(I)\varphi^{in}(I;II)\,df(I)\right\}N^{nk}(II).$$

Taking into account 3.8 we obtain from the last system of equations a system of differential equations

(5.11)
$$\frac{\partial w^{k}(II)}{\partial x^{3}}_{II} = \sum_{n=1}^{3}\sum_{\alpha=1}^{2}\sum_{\beta=1}^{2} \frac{\partial^{2}}{\partial x^{\alpha}\partial x^{\beta}}_{II\ II_{3}}[w^{n}(II)B^{nk}_{\alpha\beta}(II)] +$$

$$-\sum_{n=1}^{3}\sum_{\alpha=1}^{2}\frac{\partial}{\partial x^{\alpha}}_{II}[w^{n}(II)A^{nk}_{\alpha}(II)] + \sum_{n=1}^{3} w^{n}(II)N^{nk}(II)$$

$$\text{for } k = 1,2,3.$$

The system of equations 5.11 contains the functional coefficients $B^{nk}_{\alpha\beta}$, $A^{nk}_{\alpha}$, $N^{nk}$ for $n,k = 1,2,3$; $\alpha,\beta = 1,2$ determining the shaping process of the field of displacements $w^{k}$. The total number of these coefficients is

$$3\cdot3\cdot2\cdot2 + 3\cdot3\cdot2 + 3\cdot3 = 63.$$

For physical interpretation of these coefficients

let us consider a special two-dimensional plane field of displacements in relation to the coordinate system $x^1$, $x^3$. We
shall confine ourselves to only one vertical component of displacement $w^3$. In this case, the system of equations 5.11 is
reduced to a single equation containing only this component
of displacement. Component $w^3$ will henceforth be designated by
$w = w^3$. In addition, we assume that the medium is horizontal
ly homogeneous that is coefficients $B_{\alpha\beta}^{nk}$, $A_{\alpha}^{nk}$ and $N^{nk}$ de
pend only on the height coordinate $x^3$.

In that case, the system of equations 5.11 is re
duced to the equation

$$\frac{\partial w(x^1 x^3)}{\partial x^3} = B(x^3)\frac{\partial^2 w}{\partial (x^1)^2} + A(x^3)\frac{\partial w}{\partial x^1} + N(x^3)w. \qquad (5.12)$$

Let us assume the initial condition for $x^3 = 0$ in
the form

$$w(x^1, 0) = \gamma \delta(x^1 - x_0^1) \qquad (5.13)$$

where $\delta = \delta(x^1 - x_0^1)$ is the so-called Dirac function.

Solution of equation 5.12 for the initial condi
tion 5.13 takes the form

$$w(x^1, x^3) = \gamma \left[ 4\pi z(x^3) \right]^{-\frac{1}{2}} \exp\left\{ -\frac{\left[ x^1 - x_0^1 + \varrho(x^3) \right]^2}{4z(x^3)} + \alpha(x^3) \right\} \qquad (5.14)$$

where

$$(5.15)\, \mathfrak{z}(x^3) = \int_0^{x^3} B(s)ds \; ; \quad \varrho(x^3) = \int_0^{x^3} A(s)ds \; ; \quad \alpha(x^3) = \int_0^{x^3} N(s)ds \; .$$

Solution 5.14 provides interpretation of the coefficients $A$, $B$ and $N$ in equation 5.12, and consequently also functions $\mathfrak{z}$, $\varrho$ and $\alpha$ in equation 5.14

For an established value of $x^3$, this solution illustrates a Gaussian curve with its extremity in point $x^1 =$ $= x_0^1 - \varrho(x^3)$. The functional coefficient $\mathfrak{z} = \mathfrak{z}(x^3)$ connected by equation 5.15 with coefficient B in 5.12 characterizes the degree of dispersion of the Gaussian trough. In this way, for a given value of $x^3$, the smaller $\mathfrak{z}$, the larger the part of the surface lim ited by the Gaussian curve and straight line $x^3 = const.$ are concen trated around the straight line $x^1 = x_0^1 - \varrho(x^3)$.

The phenomenon of dispersion of the trough may be interpreted as a result of unordered random walk of void spaces elicited by the creation of a void at the level $x^3 = 0$ under boundary condition 5.13. In the limits for $\mathfrak{z} = 0$, the whole surface, according to boundary condition, is concentrated in the point $x^1 = x_0^1$ .

The functional coefficient $\varrho = \varrho(x^3)$ equation 5.15 is linked by this equation with coefficient $A$ in eq. 5.12 and characterizes the magnitude of deviation of the extreme point of the Gaussian trough from the vertical axis $x^1 = x_0^1$ on which the boundary condition $x^3 = 0$ was imposed in the form to the Dirac function.

The magnitude of this deviation characterizes the ordered motion of void spaces connected with the structure of the medium, i.e., anisotropy caused by sloped stratification.

With the purpose of interpretation of the coefficient $\alpha = \alpha(x^3)$ in equation 5.15 connected with equation 5.12, the equation is integrated according to the variable

$$F(x^3) = \int_{-\infty}^{+\infty} w(x^1, x^3) dx^1 =$$

$$= e^{\alpha(x^3)} \int_{-\infty}^{+\infty} \gamma [4\pi z(x^3)]^{-\frac{1}{2}} \exp\left[-\frac{[x^1 - x_0^1 + \varrho(x^3)]^2}{4z(x^3)}\right] dx^1 . \qquad (5.16)$$

Integration gives

$$F(x^3) = \gamma e^{\alpha(x^3)} . \qquad (5.17)$$

where $F(0) = \gamma$, according to the equations 5.15 $\alpha(0) = 0$.

The sense of the magnitude $F = F(x^3)$ determined by equation 5.17 is as follows. It is the total surface of the trough at the level $x^3$. The equation 5.17 gives the relation of this surface to the coordinate $x^3$. In particular, if $N(x^3) \equiv 0$ then according to equation 5.15 $\alpha(x^3) \equiv 0$, and $F(x^3) = \text{const}$, i.e. the medium is integrally noncompressible.

Hence, the functional coefficient $N = N(x^3)$ and the coefficient $\alpha = \alpha(x^3)$ determined by it, characterize the integral compressibility of the medium.

It follows that the process of formation of the

trough, described by equation 5.12 is characterized by three
functional coefficients $B = B(x^3)$, $A = A(x^3)$ and $N = N(x^3)$ .
The physical interpretation of these coefficients has been
given above.

In equation 5.12 coefficients $B$ , $A$ , $N$
are dependent only on the coordinate $x^3$ and there is no difficul
ty in obtaining an effective solution of equation 5.12 which
should appear if the coefficients of that equation depend on
$x^1$ However, a solution can be obtained by applying, for instance,
an electrical analogue of parabolic equations. Some of these solutions are illustrated in the figures. [7]

The figures 13, 14, 15, 16 represent solution of
the basic equation 5.12 in which coefficients $B$ , $A$ , $N$ are dis
continuously dependent on coordinate $x^1$ .

The figures represent the course of the solution
as functions of $x^1$ at established values of $x^3$ , the stepped
course of variable values of $B$ , $A$ and $N$ , and the schematic
structure of the medium, i.e. its stratification corresponding
to stepwise anisotropy which lacks homogeneity.

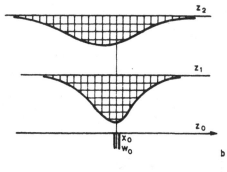

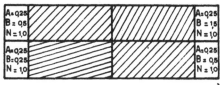

Fig. 13

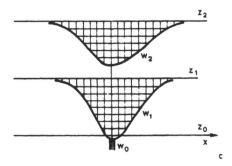

Fig. 14

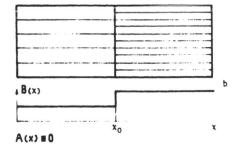

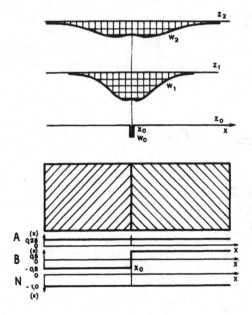

Fig. 15

Fig. 16

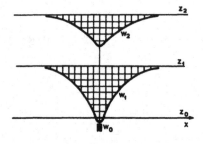

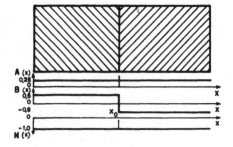

# 6. SPECIAL CASES OF SOLUTIONS OF THE SYSTEM OF EQUATIONS 5.11

We turn now to the general case of a field of displacements described by the system of linear equations 5.11.

In this case the process of displacements is described by coefficients $B_{\alpha\beta}^{nk}$, $A_{\alpha}^{nk}$, $N^{nk}$. The system of equations 5.11 includes the general case when the medium is char acterized by anisotropy, which lacks homogeneity and is describ ed by this set of coefficients, which may be functions of spatial coordinates.

The process of displacement alters the properties of the placed medium, leading to delinearization of the model describing the displacement. The coefficients describing the properties of the medium are dependent on the components of the displacement, and the phenomenon is described by a system of non-linear equations. For the present, this problem will be set aside.

Additional assumption of properties of the medium reduces the number of functional coefficients in the system of linear equations 5.10 or 5.11.

From the standpoint of experimental studies and of the possibility of comparing the results with theory, the case of a plane state of displacements in an isotropic, horizontally homogeneous medium is of special interest.

In this case, as a result of existing symmetry, as shown by Smolarski, the system of equations 5.11 is reduced to the following linear equations:

$$\frac{\partial w^1}{\partial x^3} = \frac{\partial^2}{\partial (x^1)^2}\left[B_{11}^{11} w^1\right] - \frac{\partial}{\partial x^1}\left[A_1^{13} w^3\right] + N^{11} w^1$$

(6.1)
$$\frac{\partial w^2}{\partial x^3} = \frac{\partial^2}{\partial (x^1)^2}\left[B_{11}^{11} w^2\right] + N^{11} w^2$$

$$\frac{\partial w^3}{\partial x^3} = \frac{\partial^2}{\partial (x^1)^2}\left[B_{11}^{33} w^3\right] - \frac{\partial}{\partial x^1}\left[A_1^{31} w^1\right] + N^{33} w^3.$$

Here, the process of displacements is defined by 6 functional coefficients $B_{11}^{11}, B_{11}^{33}, A_1^{13}, A_1^{31}, N^{11}, N^{33}$ .

As a result of the assumed horizontal homogeneity, the functional coefficients in the system of equations 6.1 will depend on the variable $x^3$ only.

The solution of the system of equations 5.11, and in particular 6.1 can be obtained by applying Fourier's transformation.

By employing this method, Smolarski [4] has obtained a series of solutions for particular initial conditions of essential importance in experimental studies.

These solutions for various initial conditions are illustrated in the following figures.

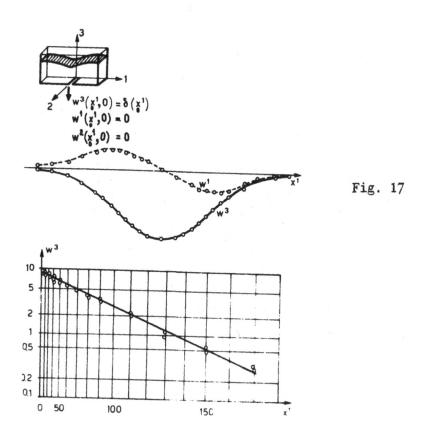

Fig. 17

Fig. 17 illustrates the solution of the system of equations 6.1 for the initial conditions

$$w^1_0(x^1,0) = 0$$

$$w^2_0(x^1,0) = 0$$

(6.2a)

(6.2b)                                  $$w^3_0(x^1,0) = \delta_0(x^1) .$$

The accompanying sketch illustrates the conditions 6.2.

In this figure, the values $w^1$ and $w^2$ measured at $x^3$ are presented in the linear scale and in the functional scale in which the solution of $w^3$ is transformed into half-straight.

The solutions of $w^1$ and $w^3$ of the system of the equations 6.1 for initial conditions 6.2 fulfil the relation:

$$\frac{w^1}{w^3} = 2\sqrt{\frac{A^{13}_1}{A^{31}_1}}\, e^{(N^{11}-N^{33})x^3}\, th\frac{\gamma}{2B}x^1$$

where     $B = B^{11}_{11} = B^{33}_{11}$   and   $\gamma = \sqrt{A^{13}_1 A^{31}_1}$ .

It follows from the last equation that at given values of $x^3$ the course of quotient $\dfrac{w^1}{w^3}$ is described by the function $th$ .

This conclusion is compared with the results of experimental measurements in Fig. 18.

Fig. 19 gives the solution of the system 6.1 for initial conditions

$$w^1_0(x^1,0) = 0$$

(6.3)                                  $$w^2_0(x^1,0) = 0$$

$$w^3_0(x^1,0) = H(x^1)w^3_{max}, \quad w^3_{max} = const.$$

where $H=H(x)$ is Heaviside's function. The initial condition 6.3 is illustrated in Fig.19 in which solutions $w^1$ and $w^3$ are represented in functional scales in such a way that the solutions

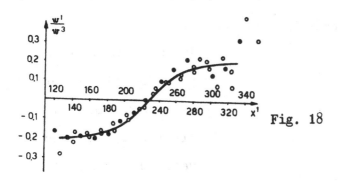

Fig. 18

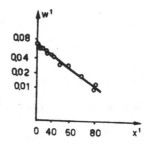

$$w^3\left(\underset{\bullet}{x^1},0\right)=H\left(\underset{\bullet}{x}\right),\ w^1\left(\underset{\bullet}{x^1},0\right)=0,\ w^2\left(\underset{\bullet}{x^1},0\right)=0$$

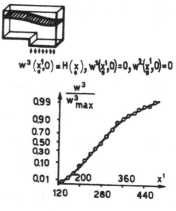

Fig. 19

are transformed into straight lines.

The course of the diagram $w^3/w^3_{max}$ deviates from a theoretical straight line. The course of $w^1$ is represented in the functional scale in Fig. 19.

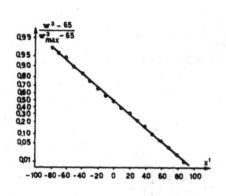

Fig. 20

It is of interest to compare the solution of system 6.1 for initial conditions 6.3 with geodetic measurements in nature, which pertain to the vertical components $w^3$ of displacements of the earth's crust caused by subterranean mining operations. The conditions of the exploitation corresponded approximately to the initial conditions described by equation 6.3. The diagram of the solution of $w^3$ at height $x^3$ corresponding to initial conditions 6.3 is plotted on a functional scale (Fig. 20) which transforms the solution into a straight line. The points around the straight line represent the results of the aforementioned geodetic measurements.

The next experiment performed with the purpose of comparing the solution of the system of equations 6.1 with measurements of displacements in a granular medium had the following initial conditions

$$w^1_0(x^1,0) = 0$$

$$w^2_0(x^1,0) = H(x^1)$$                    (6.4)

$$w^3_0(x^1,0) = 0$$

The condition is represented in Fig. 21, which also illustrates the curves obtained by measurement for initial conditions 6.4. As can be seen, the component $w^1$ differs from zero, and although its value compared with $w^3$ is very small (max 5%), nevertheless it should in theory be equal to zero.

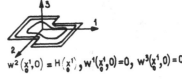

$$w^2(x^1_0,0) = H(x^1), w^1(x^1_0,0)=0, w^3(x^1_0,0)=0$$

The diagram function $w^2$ is presented in the linear scale and then plotted on the func-tional scale, which trans-forms the theoretical so-lution of $w^2$ into a straight line. The actual image of this diagram deviates from the theoretical straight line, presumably as a re-sult of nonlinear effects

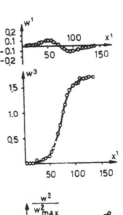

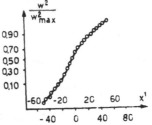

Fig. 21

in the displacement.

The next experiment designed to verify the theo‐
retical concept pertains to an axial–symmetrical field of dis‐
placements in a granular medium, which was accomplished as fol‐
lows  Fig. 22 [8] .

The granular medium was placed in a box with a
horizontal base, and the displacements were measured on its up‐
per surface. Displacements of the granular medium in the box are
caused by rotation of the disk a‐round its vertical axis by angle $\omega$ produced displacement in the gran‐ular medium above the

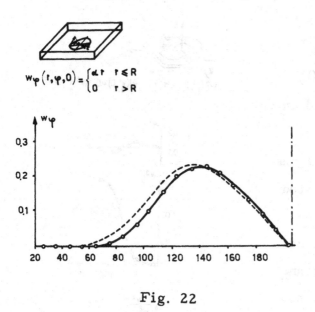

$$w_\varphi(r,\varphi,0) = \begin{cases} \alpha\, r & r \leqslant R \\ 0 & r > R \end{cases}$$

Fig. 22

disk giving rise to an axial–symmetrical field of displacements
in the granular medium in the box.

The system of equations 6.1 in the case of axial
symmetry in a cylindrical system of coordinates $\{r, z, \varphi\}$ where
$z$    is the vertical axis, $r$  the radius, and $\varphi$  the angle, as‐
sumes the form

$$\frac{\partial w_r}{\partial z} = B_{11}^{11} \frac{\partial^2 w_r}{\partial r^2} + B_{11}^{11} \frac{1}{r} \frac{\partial w_r}{\partial r} + w_r \left( N^{11} - \frac{B_{11}^{11}}{r^2} \right) - A_1^{13} \frac{\partial w_z}{\partial r}$$

$$\frac{\partial w_\varphi}{\partial z} = B_{11}^{11} \left( \frac{\partial^2 w_\varphi}{\partial r^2} + \frac{1}{r} \frac{\partial w_\varphi}{\partial r} - \frac{w_\varphi}{r^2} \right) + N^{11} w_\varphi \qquad (6.5)$$

$$\frac{\partial w_z}{\partial z} = B_{11}^{33} \left( \frac{\partial^2 w_z}{\partial r^2} + \frac{1}{r} \frac{\partial w_z}{\partial r} \right) - A_1^{31} \left( \frac{\partial w_r}{\partial r} + \frac{w_r}{r} \right) + N^{33} w_z$$

where $w_r$, $w_z$ and $w_\varphi$ are components of displacements in the cyl
indrical system of coordinates $\{r, z, \varphi\}$.

The second equation in the system 6.5 contains
only the component $w_\varphi$ . The solution of this equation was
obtained for boundary initial conditions corresponding to the
conditions of the experiment, in the form

$$w_\varphi(a, \varphi, z) = 0$$

$$w_\varphi(r, \varphi, 0) = \begin{cases} \omega r & \text{for} \quad 0 \leqslant r \leqslant R \\ 0 & \text{for} \quad r > R \end{cases} \qquad (6.6)$$

where $r = R$ is the radius of the disk in the box rotated by angle
$\omega$, $a$ is the distance to the nearest vertical wall of a sufficiently large box on whose walls the components of the displacements are equal to zero.

For boundary conditions 6.6 the solution of the
second equation in system 6.5 is as follows

$$w_\varphi(r, \varphi, z) = \sum_{n=1}^{\infty} A_n \exp\left[ \int_0^z N^{11}(s)ds - \lambda_n^2 \int_0^z B_{11}^{11}(s)ds \right] I_1(\lambda_n r) \qquad (6.7)$$

where $\lambda_n$ are the roots of equation $I_1(a\lambda) = a$ , and coefficients $A_n$ are defined by the relation

$$A_n = \frac{2wa}{I_0^2(\lambda_n a)} \frac{\left(\frac{R}{a}\right)^2}{\lambda_n a} \left[\frac{2}{\lambda_n R} I_1(\lambda_n R) - I_0(\lambda_n R)\right]$$

$I_0$ and $I_1$ are Bessel's function of the first kind, of zero and first order.

The theoretical solution of equation 6.7 was compared with measurements of $w_\varphi$ on the surface of the sand in the experiment illustrated in Fig. 22.

The values of $w_\varphi = w_\varphi(r)$ obtained from equation 6.7 for a given value of $z$ were plotted on the diagram as an interrupted line, and the points obtained by measurement were connected by a continuous curve.

# 7. DELINEARIZATION OF THE LINEAR MODEL OF A GRANULAR MEDIUM

The conception described above rests on the assumption of linearity of displacements (eq. 3.5).

The experiments that were carried out directly verified this assumption, as well as the comparison of the theoretical results, representing its implication, with the results of actual measurements. The results showed that the assumption of linearity is adequate to reality only for sufficiently small displacements.

With increasing values of the displacement, the properties of the granular medium become increasingly dependent on them, and the process becomes nonlinear.

As already mentioned, deviation from the assumption of linearity and a hypothesis for the delinearization of the model at present under consideration. There are any possibilities for such a delinearization, the choice of which is based upon observations of phenomena.

A trial of delinearization corresponding to the simplest case of equation 5.12 describing the subsidence trough will be presented.

The physical interpretation of the coefficient of this equation are given in paragraph 5.

Coefficient $B$ Eq.(5.12) characterizes dispersion

of the trough connected with unordered random walk of voids
in the granular medium.

Coefficient $A$ describes deviation of the Gaussian
trough caused by ordered movement of the voids connected with
sloped stratification of the medium.

Let us now consider a horizontally stratified
medium, in which $A \equiv 0$. Applying Dirac's initial condition, we
obtain bell-shaped Gaussian troughs in this medium. As a result,
inclined layers will appear on its slopes, indicating that $A \neq 0$
The changing displacement field alters the value of the coefficients of the equation describing that field.

Coefficient $A$ in the equation describing the
formation of the trough is therefore dependent on the values
characterizing the trough. Hence, the phenomenon has a nonlinear
character.

There is a wide range of methods for the delinearization of the model and a logical example of such a process
is suggested by experiment.

Experiments have been carried out on the relation
ship between the components of shearing strain and dilatation.

The experiments were concerned with a two–dimensional flat state of displacement measured in accordance with
the Cartesian system of coordinates $\{x^1, x^3\}$. On the basis of the
measurements, the components of a shearing strain were determined

$$\mathcal{E}_{13} = \frac{\partial w^1}{\partial x^3} + \frac{\partial w^3}{\partial x^1} \tag{7.1}$$

and dilatation

$$\theta = \frac{\partial w^1}{\partial x^1} + \frac{\partial w^3}{\partial x^3}. \tag{7.2}$$

Two series of experiments were performed. In the first, displacement was produced by removing sand through a narrow slot, according to the Dirac boundary condition. This resulted in a field of displacement with the components $w^1$ and $w^3$, which were measured, and the value of $\mathcal{E}_{13}$ and $\theta$ were accordingly computed.

In the second series of experiments, the field of displacement was produced by applying boundary conditions of displacement by means of Heaviside's function. The measurements and calculations showed that the value of $\frac{\Delta w^1}{\Delta x^3}$ was one degree smaller than the value of $\frac{\Delta w^3}{\Delta x^1}$. In view of the above, the value of $\frac{\partial w^1}{\partial x^3}$ in equation 7.1 was omitted, and it was assumed that $\mathcal{E}_{13} = \frac{\partial w^3}{\partial x^1}$.

The results illustrated in Fig. 24 refer to the first series of experiments in which the field of displacement was produced by Dirac's boundary condition.

Figure 23 shows the values $\mathcal{E}_{13} = \frac{\partial w^3}{\partial x^1}$ and $\theta = \frac{\partial w^1}{\partial x^1} + \frac{\partial w^3}{\partial x^3}$ in relation to the coordinate $x^1 = x$ at a level of $x^3 = 90\,mm$.

Fig. 24 illustrates the changes of $\xi_{13}$ and $\theta$ in the event of displacement produced by boundary conditions in the form of Heaviside's function.

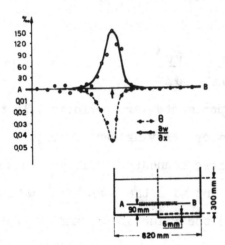

Fig. 23

When attempting to delinearize equation 5.12, its coefficients are made dependent on the value of $w^3$ characterizing the depth of the trough. The simplest hypothesis, suggested by experiment, is that coefficient $A$ in equation 5.12 will assume the form

$$(7.3) \qquad\qquad A = \alpha \frac{\partial w^3}{\partial x^1}$$

where $\alpha$ is a constant value.

Integral dilatation of the granular medium in the linear model described by equation 5.12 is characterized by coefficient $N$.

These experiments suggest that when attempting to delinearize equation 5.12, coefficient $N$ should be made dependent on the component of the tensor of the shearing strain $\xi_{13}$.

This component was assumed in the form $\varepsilon_{13} = \dfrac{\partial w^3}{\partial x^1}$ Since the magnitude of dilatation cannot depend on the sign of $\dfrac{\partial w^3}{\partial x^1}$ we assumed that it depends on $\left(\dfrac{\partial w^3}{\partial x^1}\right)^2$.

The simplest hypothesis suggested is that

$$N = \beta\left(\frac{\partial w^3}{\partial x^1}\right)^2 \qquad (7.4)$$

where $\beta$ is a constant value.

Hence, the attempt to delinearize equation 5.12, in agreement with the hypotheses 7.3, 7.4 gives, instead of a linear equation 5.12, a nonlinear equation in the form

$$\frac{\partial w^3(x^1, x^3)}{\partial x^3} = B(x^3)\frac{\partial^2 w^3}{\partial(x^1)^2} + \alpha\left(\frac{\partial w^3}{\partial x^1}\right)^2 + \beta w^3\left(\frac{\partial w^3}{\partial x^1}\right)^2. \qquad (7.5)$$

For the sake of simplicity, we continue to write $x^1 = x$, $x^3 = z$, $w^3 = w$ and we assume that $B = \text{const}$

Equation 7.5 is a special example of the following nonlinear equation [9]

$$\frac{\partial w}{\partial z} = B\frac{\partial^2 w}{\partial x^2} + f[w]\left(\frac{\partial w}{\partial x}\right)^2. \qquad (7.6)$$

In that special case, in equation 7.5 is

$$f[w] = \alpha + \beta w. \qquad (7.7)$$

Let us now turn our attention to the solution of Cauchy's problem, equation 7.6 which we propose to solve in the following way

(7.8) $$w(x,z) \;=\; w^*\big[\eth(x,z)\big]$$

assuming that function $\eth = \eth(x,z)$ fulfils equation

(7.9) $$\frac{\partial \eth(x,z)}{\partial z} \;=\; B\,\frac{\partial^2 \eth(x,z)}{\partial x^2} \;.$$

Substituting 7.8 in equation 7.6 we get

(7.10) $$\frac{dw^*}{d\eth}\left(\frac{\partial \eth}{\partial z} - B\frac{\partial^2 \eth}{\partial x^2}\right) \;=\; \left[B\frac{d^2 w^*}{d\eth^2} + f(w^*)\left(\frac{dw^*}{d\eth}\right)^2\right]\left(\frac{\partial \eth}{\partial x}\right)^2 .$$

Since, according to 7.9 the left side of equation 7.10 is equal to zero, and $\dfrac{\partial \eth}{\partial x} \neq 0$, hence

(7.11) $$B\frac{d^2 w^*}{d\eth^2} + f(w^*)\left(\frac{dw^*}{d\eth}\right)^2 \;=\; 0\;.$$

From equation 7.11 we get

$$\frac{dw^*}{d\eth} \;=\; K\exp\left[-\frac{1}{B}\int\limits_{0}^{w^*} f(s)\,ds\right]$$

where $K$ is an integral constant.

      Isolating the variables in the last equation, we get

(7.12) $$F(w^*) \;=\; \int\limits_{0}^{w^*}\exp\left[\frac{1}{B}\int\limits_{0}^{\xi} f(s)\,ds\right]d\xi \;=\; K\eth + K_1$$

where $K_1$ is a new integral constant. Since function $F = F(w^*)$ being an integral of the exponential function, is a monotonic function, it may be inverted. Hence, we get from equation 7.12

(7.13) $$w^* \;=\; \varphi(K\eth + K_1)\;.$$

The problem consists in finding the solution of equation 7.6

for initial conditions

$$w(x,0) = g(x).$$

Hence 7.12 corresponds to

$$F[g(x)] = K\delta + K_1$$

or, designating

$$\delta(x,0) = h(x)$$

and assuming constants $K=1$, $K_1=0$ we get

$$\delta(x,0) = h(x) = F[g(x)]. \qquad (7.14)$$

Having the initial condition $\delta(x,0) = h(x)$ we may construct

Cauchy's solution of equation 7.9

$$\delta(x,z) = \int_{-\infty}^{+\infty} \frac{h(s)}{2\sqrt{\pi Bz}} e^{-\frac{(x-s)^2}{4Bz}} ds. \qquad (7.15)$$

By means of this solution, on the basis of equations 7.13, 7.14,

7.15 we arrive at the desired solution

$$w(x,z) = \varphi\left\{ \int_{-\infty}^{+\infty} \frac{F[g(s)]}{2\sqrt{\pi Bz}} e^{-\frac{(x-s)^2}{4Bz}} ds \right\}.$$

For example, let us consider the case in which we

assume that function $f=f(w)$ is derived from equation 7.7 in the

form

$$f(w) = \beta w.$$

Equation 7.6 then takes the form

(7.16)
$$\frac{\partial w}{\partial z} = B\frac{\partial^2 w}{\partial x^2} + \beta w\left(\frac{\partial w}{\partial x}\right)^2$$

which corresponds to the conditions which do not take into account inclined stratification for the granular medium (e.g. by means of mica lamellae), only a change in dilatation of the medium produced by shearing strain $\varepsilon_{13}$ ).

The initial conditions in the example under consideration were assumed in the form

(7.17)
$$g(x) = \begin{cases} 0 \\ w = \text{const} \quad \text{for} \quad a \leqslant x < b . \\ 0 \end{cases}$$

Introducing the following designation

$$\frac{w}{w} = \bar{w}; \quad w\sqrt{\frac{\beta}{2B}} = \varrho; \quad \frac{x}{2\sqrt{Bz}} = \xi; \quad \frac{a}{2\sqrt{Bz}} = \xi a; \quad \frac{b}{2\sqrt{Bz}} = \xi b$$

we obtain the solution of equation 7.16 or initial conditions (7.17) in the form

$$\bar{w} = \frac{1}{\varrho}\sigma\left\{\mu(\varrho)[\Phi(\xi - \xi_b) - \Phi(\xi - \xi_a)]\right\}$$

where

(7.18)
$$\mu(\varrho) = \int_0^\varrho e^{s^2}ds$$

and $\sigma$ is the reciprocal function of $\mu$ and $\Phi$ is erof, i.e.

$$\phi(\varrho) \;=\; \frac{1}{\sqrt{\pi}} \int\limits_{0}^{\varrho} e^{-s^2} ds \;.$$

For a given height $z$ , the curves in Fig. 24 illustrate func-

tion $\bar{w}$ for

different values

of parameter $\varrho$ .

In comparison, a

case is given in

which dilatation

equals zero, i.e.

$\beta = 0$ , and $\varrho = 0$ .

The curve indicates

that with increasing

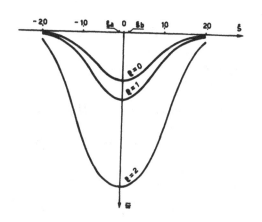

Fig. 24

$\varrho$  the volume of the trough produced by initial conditions 7.17

also increases.

       In the second example, we assume function $f = f(w)$

in the form

$$f(w) \;=\; \alpha + \beta w \;.$$

Equation 7.6 then assumes the form

$$\frac{\partial w}{\partial z} \;=\; B\frac{\partial^2 w}{\partial x^2} + (\alpha + \beta w)\left(\frac{\partial w}{\partial x}\right)^2 . \qquad (7.19)$$

This equation takes into account the process of delinearization

produced by the change in inclination of stratification during

the process of displacement (coefficient $\alpha$ ), as well as delinea

rization due to changed dilatation as a result of shearing strain
(coefficient $\beta$ ).

The following example illustrates the solution of
equation 7.19 for the following initial conditions

$$w = W = const \quad for \quad -\infty < x < 0$$

(7.20)

$$w = 0 \quad\quad\quad for \quad 0 \leqslant x < -\infty .$$

Let us consider the following 3 cases

a) $\alpha > 0, \quad \beta = 0$

b) $\alpha = 0, \quad \beta > 0$

c) $\alpha = 0, \quad \beta < 0 .$

Case a) corresponds to delinearization caused by change in the
slope of stratification of the medium, but does not take dilata
tion into account. Case b) and c), on the other hand, take into
account only change of dilatation connected with shearing strain
$\varepsilon_{13}$ .

Introducing

$$\eta = \frac{x}{2\sqrt{Bz}}; \quad \varrho = \frac{\alpha}{B}W; \quad \bar{w} = \frac{w}{W}$$

we get the solution of equation 7.19 for initial conditions 7.20
and in case a) in the form

(7.21)          $$\bar{w} = \frac{1}{\varrho} \ln\left[(1 - e^{\varrho})\Phi(\eta) + e^{\varrho}\right]$$

where

$$\Phi(\eta) \;=\; \frac{1}{\sqrt{\pi}} \int_{-\infty}^{\eta} e^{-s^2} ds \;.$$

Function 7.21 is illustrated in Fig. 25.
The figure gives the so
lution for different
values of parameters $\varrho$ .
In comparison, the
broken line for $\varrho = 0$
depicts the curve cor,
responding to the so-
lution of the linear equation

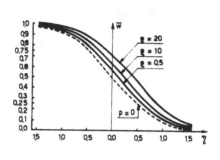

Fig. 25

$$\frac{\partial w}{\partial z} \;=\; B \frac{\partial^2 w}{\partial x^2} \;.$$

The solution of equation 7.19 in case b) for initial conditions 7.20 has the form

$$\bar{w} \;=\; \frac{1}{q}\eth\{\mu(q)[1 - \Phi(\eta)]\} \qquad\qquad (7.22)$$

where $q = \sqrt{\dfrac{\beta}{2B}}$ and the remaining designations are the same as in
the preceding example in case a) and $\mu$ as in equation 7.18.
The solution of equation 7.19 in case c) for initial conditions
has the form

$$\bar{w} \;=\; \frac{1}{r}\Omega\left\{\left[\frac{1}{2} - \Phi(r)\right]\Phi(\eta) + \Phi(r)\right\} \qquad\qquad (7.23)$$

where, besides the designations as above $\gamma = \sqrt{\dfrac{-\beta}{2B}}$ and $\Omega$ denotes the reciprocal of function

$$\phi(x) \;=\; \frac{1}{\sqrt{\pi}} \int\limits_{-\infty}^{x} e^{-s^{2}}\, ds \;.$$

The diagrams of functions 7.22 and 7.23 are illustrated in
Fig. 26 which contains, for purpose of comparison, the interrupted curve, representing the solution of the linear equation
$\dfrac{\partial w}{\partial x} = B\,\dfrac{\partial^{2} w}{\partial x^{2}}$   for initial conditions 7.20.

# REFERENCES

[1] J. Litwiniszyn: "Application of the Equation of Stochastic Processes to Mechanics of Loose Bodies" - Arch. Mech. Stos. T I, 8, 1956

[2] S. Goldstein: "On Diffusion by Discontinuous Movements, and on the Telegraph Equation" - The Quarterly Journal of Mechanics and Applied Mathematics - Vol IV, Part 2, 1951

[3] J. Litwiniszyn: "An Application of the Random Walk Argument to the Mechanics of Granular Media" - Rheology and Soil Mechanics Symposium Grenoble, April 1-8 1964, Springer-Verlag, 1966

[4] A. Smolarski: "On Some Applications of Linear Mathematical Models to the Strata Mechanics" - O/PAN w Krakowie - Prace Komisji Nauk Technicznych - Mechanika 3, 1967

[5] J. Litwiniszyn: "A Solution of the Smoluchowski Equation and the Possibility of its Application in Mechanics of Stochastic Media I, II - Bull. Acad. Pol. Sci. Ser. Tech., Vol. IX, N° 5,6 , 1961

[6] M. Ghermanescu: Acad, R.P., Rom. Bul. Sti. mat.fiz. 5 , 1953

[7] W. Trutwin: "Trough Subsidence Profiles in Unisotropic and Non-Homogeneous Stochastic Media" - Doctor Thesis, 1962

[8] R. Romanica: "Theoretical Calculation and Experimental Measurement of the Angular Component of Displacement of Loose Media with Rotational Symmetry" - Bull. Acad. Pol.Sci.Ser.Techn.16 , 1968

[9]    J. Litwiniszyn, A.Z. Smolarski:  "On a Certain Solution
       of the Equation $\psi(\tau)\upsilon'_\tau = A\upsilon''_{22} + f(\upsilon)(\upsilon'_2)^2$   and
       its Application to the Problem of Mechanics of
       Loose Media" - Bull. Acad.Pol.Sci.Ser.Techn. Vol.
       X, N° 3,  1963

# CONTENTS

Page

Preface .......................................... 3

1. Stochastic Methods in Mechanics of Granular
   Bodies ....................................... 5

2. Heuristic Models Based on the Concept of
   Random Walk .................................. 9

3. General Method of Displacements of a Granular
   Medium and its Postulates 1,4 .............. 36

4. Method of Solution of the System by Means of
   Fourier's Transformation 5,4 .............. 45

5. The Second Method of Solving the System of
   Smoluchowski Equations (3.12) 1,4 ........ 56

6. Special Cases of Solutions of the System of
   Equations 5.11 ............................ 69

7. Delinearization of the Linear Model of a
   Granular Medium ............................ 79

References ....................................... 91

Printed in the United States
By Bookmasters